Et ego in Arcadia vixi

FOOTNOTES
ON NATURE

John Kieran

WOOD ENGRAVINGS BY NORA S. UNWIN

DOUBLEDAY & COMPANY, INC., GARDEN CITY, N. Y.

1947

O'Hare

KIERAN, JOHN. Footnotes on nature; wood engravings by Nora S. Unwin. 279p $3 Doubleday

574 Nature   47-30833

Random memories of nature walks which the author has taken with a group of friends, spiced with anecdotes of Mr Kieran's boyhood on a Dutchess county farm. Most of his trips were taken in the wilds of Westchester or New England, but some were in the immediate vicinity of New York city. The appendix contains lists of the birds, flowers, trees, and insects encountered on his walks and mentioned in the book.

Booklist 44:7 S 1 '47

"Not only does this volume reveal Kieran as a writer of unusual ability, but it indicates that fundamentally he has the soul of a poet. Here we follow him in his various and meandering adventures as a nature-lover, especially as a devotee of birds and bird lore. . . And it is all done simply, without literary flourishes." John Drury

KIERAN, JOHN, 1892-

. . . Footnotes on nature; wood engravings by N. S. Unwin. Doubleday 1947 279p illus $3

A "combination of familiar essay and useful field book—and in it the author's personality bubbles to the surface again and again. He has filled his book with curious sidelights on the great men who came to walk with him in New York City's Van Cortland Park—as well as his regular walking friends—he has recalled walks of his childhood days—and, of course, he has included a number of couplets and quatrains from his favorite poets." Huntting

Footnotes on Nature

by John Kieran

FOOTNOTES ON NATURE
POEMS I REMEMBER
JOHN KIERAN'S NATURE NOTES

Foreword

This is a book about a few men and many walks in the
woods and fields of New York and New England. It is,
to a large extent, an account of the birds, trees, flowers
and other forms of life that we encountered outdoors in
that territory. Since the common names of birds, trees
and flowers vary from one district to another and even
the scientific names may be changed from time to time,
for the sake of clarity I have used only one common
name for any species mentioned in the text, and in the
Appendix, after each common name, I have listed the
scientific name in accordance with the following
authorities:

Birds: *Handbook of Birds of Eastern North
America,* by Frank M. Chapman, Second Revised Edition, Appleton-Century, 1940.

Botany: *Illustrated Flora of the Northern States
and Canada,* by Britton and Brown, Second
Edition, New York Botanical Garden, 1943.

Insects (except Butterflies and Moths): *Field Book of Insects,* by Frank E. Lutz, Third Edition, G. P. Putnam's Sons, 1935.

Butterflies and Moths: *The Butterfly Book,* by W. J. Holland, Revised Edition, Doubleday, Doran & Co., 1931; *The Moth Book,* by W. J. Holland, Doubleday, Doran & Co., 1937.

Mammals: *Mammals of America,* Technical Editor, Harold E. Anthony, Garden City Publishing Co., 1937.

The writing of this book has been a labor of love. My only regret is that in its pages I have failed to do justice to an entrancing subject and to the Corinthians, lads of mettle—good boys, by the Lord!—who were my regular companions on journeys that stretched from the sea to the hills and from the clattering city streets to the quiet woodland ways.

Footnotes on Nature

1

CHARMIAN
Is this the man? Is't you, sir, that knows things?
SOOTHSAYER
In Nature's infinite book of secrecy, a little I can read.
ANTONY AND CLEOPATRA, ACT I, SC. 2

Where it all started I do not clearly remember, but "in the dark backward and abysm of Time" the first bird note that returns to me across—*eheu, fugaces!*—the vanished years is the strident shriek of the Crested Flycatcher. When I was about eleven years old my father bought a forty-acre farm in Dutchess County, N.Y., to serve as a "Summer place" for our family that included two parents, seven children, numerous cheerful relatives and sundry pet stock such as dogs, cats, rabbits and pigeons. The farmhouse was a rambling affair of twelve rooms with low ceilings and creaking floors. There was a wonderful well on the front porch and another good well in the barnyard.

1

The farm was a fine place on which to turn children loose during the long Summer vacations and it wasn't long before we had explored the whole countryside. Just beyond the barn there was an ancient orchard and there, in the hollow limb of a decrepit apple tree, a Crested Flycatcher had built its nest. Few birds advertise their presence more loudly than this species. I heard the bird during our first Summer on the farm and, by watching it, soon discovered its nest. When the young birds had grown up and wandered off, I looked into the empty nest and found the cast snakeskin that is usually twisted loosely into the fabric of a Crested Flycatcher's nest, presumably for the purpose of frightening off predators.

Our first year on the farm was the last year that saw Bob-whites or Quail abundant in that territory. The following Winter was so severe that almost all the Bob-whites were killed off and never again did I see or hear one in our valley. But during that first Summer we heard them calling "Bob White! Oh, Bob White!" all around the dooryard and orchard, and in our rambles through the fields we often flushed a bevy that sputtered up hastily from around our feet and scattered in flight to all points of the compass.

So the Crested Flycatcher and the Bob-white are officially identified with our first Summer on the farm, yet I must have known a few birds at an earlier period. I was born in the suburbs of New York City—Kingsbridge, to be exact—in a house that was surrounded by a lawn and a hedge and a few trees. There were Robins

on the lawn as far back as I can remember. And certainly I knew Crows when I was a small child. In those days the street lamps burned gas and the lamplighter making his rounds at dusk was as regular a sight as the postman with his bag or the policeman swinging his stick on patrol. My mother usually had afternoon tea in an upstairs room that faced westward. Tea and twilight arrived together on short Winter days and I can recall standing by the window at teatime and watching, as the lamplighter worked his way up the snow-covered street, a long flight of Crows winging slantwise across the sky in the wintry dusk.

Surely I must have known the Baltimore Oriole when I was in knee pants and certainly I recognized a Bluebird at close range, but beyond these abundant species and a few general identifications such as "sparrow" or "hawk" or "seagull," I was blindly ignorant of the variety of birdlife around me. The first real suspicion that there might be more birds about than were dreamt of in my philosophy came in one of my later Dutchess County

3

Summers when I was walking with some boys and girls across the pasture land of a neighboring farmer, an elderly and stout Irishman of genial disposition who was known all over the countryside as Pat. This Pat was a good man but a poor farmer and preferred to do odd jobs for Summer residents rather than stay at home and farm his own land. He kept a small herd of dairy cows in a vague way.

Pasture land makes easy walking. The cows and horses pave the way, so to speak. They keep the grass cropped short and they make paths through the gullies and along the steep hillsides. This day in Pat's pasture we saw a small bird, yellow and black, swing past us in nonchalant flight, singing sweetly as it went. A girl in the party asked me the name of the bird. In absolute ignorance but with an air of utmost assurance, I blithely answered: "Vireo." This was a bird name that I had come across by chance in some pure reading matter and it made an impression on me. I kept it in mind and tossed it off at the first chance I had to impress somebody else with it. The bird was, of course, a Goldfinch, though I did not learn that until much later. But what sticks in memory apropos of the incident is the secret sense of guilt that I felt because I didn't know what the bird was and had thrown an outrageous bluff to cover my ignorance. I wore a heavy sprinkling of freckles all over my face through my growing years and I knew that freckles were supposed to be a guarantee of honest character beneath, but I was a complete fraud in this bird case. Though nobody in the group caught me at it,

4

I decided to lead a better life. Since I loved the outdoors so much, I certainly should know more than I did about birds and trees and flowers.

I did, indeed, love the country "from childhood's earliest hour." As a boy I haunted the Bronx Zoo and came to feel a personal acquaintance with many of the caged animals. I read Nature books and doted on animal stories, including some by authors who were roundly denounced by the then President of the United States, Theodore Roosevelt, as "Nature fakers." I kept Rabbits behind the stable and raised pigeons in the stable loft. In our Summers on the farm we had a horse and a cow. I fed, watered and curried the horse. I milked the cow. We always had dogs of our own and somehow I managed to round up all the stray dogs in our neighborhood to add to our collection. Every so often a dignified citizen would come around and reclaim his dog, casting an accusing eye on me in the process.

During our Summers in Dutchess County I was rarely under a roof from June until September except for necessary attendance at such vital meetings as the family breakfasts, dinners and suppers (and cookies and milk before going to bed) at the family table in the farmhouse. I roamed the fields and the woods the rest of the day and at night I slept in a tent that was pitched on a little hill north of the farmhouse. On clear nights I would move my cot out of the tent and sleep on the open hillside under the moon and the stars. I knew every foot of the territory for miles around. I could

5

follow old wood roads and I could lead the way by short cuts across the fields. I had marked the locations of clumps of Trailing Arbutus, Mountain Laurel and Pink Azalea. I found wonderful patches of long-stemmed wild strawberries (could they have been *Fragaria vesca?*) of astonishing size and luscious flavor. I knew where the best holes for trout were along the brook in the woods and I knew the best stands on the hills in the hunting season—for I was a hunter in those days and, like Chaucer's Priest in the *Canterbury Tales*—

> . . . *yaf nat of that text a pulled hen*
> *That seith that hunters been nat holy men.*

But for all that, it was astonishing how little I knew about birds. And humiliating too. Driven by a guilty conscience in the matter of the miscalled "Vireo," I did learn a little—a very little—about birds in the next few years, chiefly in my Summer ramblings by hill and dale in Dutchess County. I saw and recognized a Great Blue Heron as it was standing in a brook that meandered through a meadow. There was a Phoebe that nested year after year on one of the beams that held up the roof over our front porch. House Wrens nested in a knothole on the north side of the house. With some companions I found the nest of an Eastern Green Heron in the willows surrounding a scum-covered pond in a hollow in our back fields. There were two young birds in the nest, fairly well grown but not quite able to lift themselves into the air and escape from us when we descended upon them. We took the young birds home

6

with us and tethered them with string. A few hours later the young birds had disappeared. They simply pecked at their tethers until the strands parted, and then they walked off.

We had on our front lawn a huge old Forsythia bush that flourished out of all reasonable dimensions. It was a thicket in itself. Once during the dinner hour—noon in New England, if you please—those on the right side of the table could see through the screen door and the windows that a Ruffed Grouse had flown into the shade of the Forsythia and was standing there motionless. We watched it until it walked slowly under the big bush and disappeared in the welter of trailing branchlets. A month later when I was coming home from a fishing trip to the woods of a rainy afternoon I was startled to find something brown rushing along the ground at me in a field that was dotted with scrub growth and covered with low blackberry vines. I pulled up short and saw that it was a Ruffed Grouse, with all its feathers puffed out, making the threatening gesture. When I stood still, the bird flopped away with one wing trailing as though it were broken. From the Nature stories I had read, I knew what was up. I walked forward cautiously to where the mother bird had started its rush for me and, sure enough, young birds not much larger than sparrows began to pop up from the grass and flutter off in different directions. I withdrew quietly to allow the worried mother to collect the chicks again, but I was quite stirred by this adventure and reported it enthusiastically to my family. They took it calmly.

7

One more acquaintance that I made was a yellowish bird dotted with black that I used to see on telegraph poles and fence rails. I suspected from its shape and habits that it was a woodpecker of some kind, but I had no name for it until the young farmer who lived "up the road a piece" told me it was a "High-hole," which name served me until I later learned that it was called the Flicker (or Golden-winged Woodpecker) in more formal circles. Then there was the sound in the night, the bird that I knew only by ear, a queer thing altogether while the mystery lasted. Out of the sky in the dark would come these strange noises from invisible birds passing overhead. As I lay on my cot in the tent I thought them as hoarse as the Raven that croaked the fatal entrance of Duncan under the battlements of Macbeth. Once again I turned to the young farmer up the road for an explanation and he said the invisible birds were "Quowks," but it was some years later in Van Cortlandt Park swamp in New York City that I first laid eyes on Black-crowned Night Herons and recognized them as the owners and operators of those weird voices heard under cover of darkness in Dutchess County.

There were other sounds to be heard in the eerie hush of Summer nights. At midnight in my unguarded tent I once heard what seemed like the sound of slow footfalls in the dead leaves that carpeted the ground in a Black Locust grove near my tent. Certainly something was walking or moving about at short intervals there in the darkness. I slipped quietly from my cot to investigate. As I approached it, the sound ceased. When I went

8

off a bit, the sound of half-shuffling footsteps began again. Ultimately I tracked down the culprit. It was a toad on a night prowl. One season we had stunning sound effects in our little valley. A family of Red Foxes grew up on the hillside opposite my tent and they used to "squall" at night around an old red barn, probably because there were chickens and turkeys roosting in the barn and on the low branches of trees around it. This "squalling" is unearthly—almost demoniac—in quality. I wish I were sure that the vixen "squalled" so that I could filch a vividly descriptive line from Tennyson and say that I heard

the shrill-edged shriek of a mother divide the shuddering night.

But best of all Summer sounds I loved the baying of the hounds on the trail of some Jackrabbit or fox in the night. Many times, on hearing the hounds running in the moonlight, I would dress myself, no matter what the hour, and go off to be near the chase. Often I sat for hours on a hilltop listening to the music of the hounds floating up from the dim valley in the deep quiet of a moonlit night. With this youthful background, I later came to love that passage in *Midsummer-Night's Dream* where Hippolyta says:

> *I was with Hercules and Cadmus once*
> *When in a wood of Crete they bay'd the bear*
> *With hounds of Sparta; never did I hear*
> *Such gallant chiding; for, beside the groves,*
> *The skies, the fountains, every region near*
> *Seemed all one mutual cry; I never heard*
> *So musical a discord, such sweet thunder.*

9

Such, by and large, were my sentiments when in June, 1912, I was graduated from college and moved on to earn a living as best I could. I did not want to work in New York City. It was too crowded. I hated paved streets, packed subway trains and vast areas of bustling business buildings or smelly tenements. I loved the wooded hills, the hayfields, the apple orchards and the little brooks of Dutchess County. So I went back to the farm to raise chickens. An old Irishwoman who kept a neighborhood store at which we traded in the city heard of my decision and said with the utmost contempt:

"Raisin' chickens, is it? After all his damn schoolin'!"

Nevertheless, it was a good Summer and a glorious Autumn for me. I built chicken coops all through the Summer in preparation for rearing a big flock the following Spring. In September I picked about two hundred barrels of apples and perhaps a dozen barrels of pears and shipped them off to market. What finer occupation is there in the wide world than that of picking apples of a bright Autumn day with a light west wind blowing and scattered white clouds floating fleecily across a deep blue sky? This is not just a reminiscent dream with me. I knew at the time and so said to myself that it was simply wonderful to be twenty or thirty feet up an extension ladder (our old trees had been allowed to grow too high) picking deep red Baldwin apples in the sparkling sunlit atmosphere, with the landscape all around set in gorgeous Autumn colors. It is no wonder that I sang at my work as the ladder swayed and the

10

apple tree branches and leaves rustled in the cool breeze out of the northwest. I thought then with Bryant—such thoughts really did come to me, and still do—

> *Ah, 'twere a lot too blest*
> *Forever in thy colored shades to stray;*
> *Amid the kisses of the soft Southwest*
> *To roam and dream for aye;*
>
> *And leave the vain low strife*
> *That makes men mad—the tug for wealth and power—*
> *The passions and the cares that wither life*
> *And waste its little hour.*

For family driving that Summer we had a four-year-old colt, a lively sorrel with a blazed face. When the family returned to the city at the end of the vacation period, I broke the colt to saddle. On bright September days I rode across the harvest fields and down the colored archways of the country lanes, knowing that it was good for me to be there. At twilight there were Sugar Maples that seemed to be lighted from within by some mysterious rose-purple glow that came through their red-and-yellow leaves. I felt the beauty of such surroundings to whatever depths were in me, and I still recall the very spot on the wood road where, in the thrill of one of those Autumn rides, I said to myself: "This is surpassingly beautiful and I love it. Never in the future can I be any happier than I am now." Possibly I never have been happier; it's difficult to bring such joyous matters to sober measurement.

Out of a clear September sky I was asked by the chairman of the District School Board to undertake the

11

task of teaching the six pupils who would attend the district school that year. The school building was a little unpainted shack in a fringe of woods just east of the railroad cut above Anson's Crossing, which was a "whistle stop" on the Newburgh, Dutchess & Connecticut Railroad, now defunct, not "its irised ceiling rent, its sunless crypt unsealed," but its rails and ties removed and its right of way returned to the wild state. Under the shingled roof of the shack that served as a school there was an entrance hallway that was used to hold hats and coats and to store firewood for the chunk stove that stood in the middle of the schoolroom into which the hallway opened. There were half a dozen rows of desks for pupils, one raised desk at the front for the teacher, and a blackboard on the front wall.

By virtue of being a college graduate, I was allowed a "temporary license" to teach there, and I went to work at it for the sum of forty dollars a month, less 1% for some retirement fund of which I never took advantage. It was then that the hand of the law was laid upon me, and I was compelled by "statoot made an' perwided," in the words of Mr. Weller, to buckle down and really learn something about birds. I discovered that rural teachers in that area had to give a Nature Study Course in a small way to their pupils. Part of this program in my term of office included teaching the pupils to recognize four common birds of the region. Colored pictures of the chosen birds, with some pure reading matter attached, were provided by the Department of Education. When it came time to teach this subject, I picked up the

12

first leaflet and saw on it a picture of an odd-looking bird in what seemed to me to be an utterly impossible position. It was a stumpy-tailed bird about six inches long, white underneath, gray and black on top, and it was pictured apparently going down an old fence post headfirst. I never had seen any bird proceed in that topsy-turvy fashion and, furthermore, the bird of the picture was a total stranger to me. I glanced at the pure reading matter under the picture and it ran something like this:

"The White-breasted Nuthatch. This common bird is known to every farm boy and girl. . . ."!

I looked at the picture again in astonishment. No, sir; never before in my life had I seen anything that looked like that bird, and I had been outdoors in that area for a dozen Summers and many weeks at other seasons of the year. Not only that, but the confounded bird was shown *walking down* a fence post, a most irregular procedure in my view. There was no lesson in Nature Study that day. The bird leaflet went quietly back into the drawer of teacher's desk and I took up some subject I could handle with greater confidence: spelling. I decided that the bird problem could go over until the next morning. I would sleep on the mystery. *La nuit porte conseille.*

As usual, I slept outdoors. My cot was on an open porch facing the east. It was October, with consequent cool weather, and the next morning when I awakened I lingered under the blankets a few minutes before getting up. About ten feet away on the lawn there was a

13

cultivated Black Cherry tree. As I lay there I noticed something moving on the trunk of the tree. The moving object was, to my utter amazement, the mysterious bird "known to every farm boy and girl," the aforesaid White-breasted Nuthatch of the Nature Study leaflet— *and it was moving down the tree headfirst!* I reared up on my cot to have a better look at this phenomenon and my sudden movement caught the bird's attention so that it paused in its downward journey to twist its head to stare at me, which put it momentarily in the exact pose of the bird in the picture that the Department of Education had forced upon me.

This experience was a stunner and gave me furiously to think. I never had suspected the existence of any such bird until I had seen its picture and a few printed words about it the previous day. But the first thing that met my eyes the next morning was a live specimen of this type not ten feet from the end of my nose! I decided to begin to look into the matter immediately. It was nearly a mile from our farmhouse to the school and, on the way that morning, I kept my eyes open with astonishing results. I saw four more of these birds going up or down the trunks of trees! By the time I reached the school I realized that I had been practically blind for twenty years. While looking for White-breasted Nuthatches on this morning walk, I noted that there were odd-looking sparrows in the undergrowth. They had white lines on the sides of their heads and white patches under their throats. That was my introduction to the White-throated Sparrow, though I did not know its

14

name. I saw other birds I never had noticed before, but what they were I couldn't guess. I determined to keep my eyes open and investigate further.

It was, of course, the wrong time of year for any person to attempt a primary course in bird study in the New England area, which this part of New York is essentially. The migration was largely over. Most of the warblers had departed and what few were still passing through were in sober Autumn plumage. Even the sparrows are duller at this time of year and the young in their first traveling clothes add to the confusion. However, we did put bread crumbs on the window sill at the school and Black-capped Chickadees soon became regular boarders at our table. One of these friendly little birds, seeking an extra-early breakfast, came through a knothole in the shutter before school was opened and was trapped between the shutter and the window. We caught the bird and the six children had a close look at it before we turned it loose again. Aside from that, I don't recall any other bird acquaintances I picked up through the Winter and I haven't the faintest recollection of the three birds other than the White-breasted Nuthatch that were part and parcel of the bird-recognition program for pupils that year. But I did keep an eye out for birds when Spring came back to Dutchess County and I remember the pleasure with which I saw the first Baltimore Oriole of the season in the top of an American Elm.

Later in the Spring I had a real find. Going home from school one afternoon, I saw a tiny bird of striking colors and odd pattern in the undergrowth along an old stone wall. The bird was a mixture of green, brown, reddish-chestnut, white and gold. The gold came in a patch on top of its head. I was entranced by the bird. When I saw it there three or four afternoons in succession, I suspected that it had a nest in the vicinity, so I went on my hands and knees among the bushes and found the nest—with four eggs, I think—just a few feet above the ground. Later, after I had purchased Chapman's *Handbook of Birds of Eastern North America,* I learned that my beautiful little bird was the Chestnut-sided Warbler. I still think it is a lovely thing to see. But in my early acquaintance I had no name for it, nor could I guess the identity of a little fluttering apparition in black and orange that zigzagged through the Spring greenery one day in May when I took the children for a walk along the brook in the woods. It was, as I later learned, the Redstart, an inadequate name in English, but allegedly derived from *Rothstert,* the German for "Red Tail," which is a little more appropriate, though it still doesn't cover the flashing picture. Edward Howe Forbush, in *Birds of Massachusetts,* is authority for the statement that the solution is more fitting in Spanish, presumably in the Latin-American countries where the birds spend the Winters. Forbush comments on the fact that the name for warblers as a group in Spanish is *Mariposas,* meaning "Butterflies," but for the Redstart in particular the Spanish name is *Candelita,* which means "Firefly." I call that wonderful and, in view of the uncouth names that have been affixed to some of our

16

beautiful birds, perhaps the general situation could be improved by a few translations from the Spanish.

My farmer friend up the road had a touch of sporting blood in him. He owned a couple of fine shotguns, a succession of handsome English Setters and a long dynasty of Foxhounds, but the apple of his eye was a beautiful chestnut filly named Varvana that he had purchased at an "Old Glory" sale at the original Madison Square Garden in New York City and that had grown up to a trotting mark of 2:27, which was impressive in our little circle. He sent her off occasionally to compete on the Grand Circuit with the aristocracy of harness racing. When the evening chores were finished he would take Varvana out of her special box stall and let her graze in a meadow below the house. A little trout brook ran audibly through the meadow and it was pleasant to hear it as background music when he and I sat there in the twilight keeping close watch on Varvana grazing around us.

"Sunset and evening star" always have had an eerie lure for me. As Jessica was never merry when she heard sweet music, twilight settling over a landscape casts a spell on me, a vague melancholy that brings on

> *A feeling of sadness and longing*
> *That is not akin to pain,*
> *And resembles sorrow only*
> *As the mist resembles the rain.*

As we sat there on long Summer evenings with twilight slowly deepening into dusk, a Whip-poor-will

17

came regularly to sit on a smooth stone quite close to us and bob its head as it uttered its vibrant call: "Whip-poor-WILL! Whip-poor-WILL! Whip-poor-WILL!" There was a wild, forlorn, plaintive touch to it that made such a lasting impression on me that if I hear the bird now in the dusk, I am a boy again sitting in a meadow at twilight with my farmer friend, the filly grazing nearby, and the Whip-poor-will is calling to me out of the mist of vanished years. "So sad, so strange, the days that are no more."

Just one more early introduction among bird acquaintances and I'll get on to less important matters. A great playground for all of us on the farm was a "Pine Forest" of some six or seven acres with a brook running along the edge of it. The trees were young White Pines from fifteen to thirty feet tall and we used to shinny up the trees and swing from one tree to the other, or sway down to the ground on the bending boughs, accumulating a monstrous amount of resin on our hands and our clothes in the process. It took much laborious scrubbing to clean our hands for dinner, but we did love the "Pine Forest"—the fragrant branches through which we swung and the brown carpet of dead needles that made such a soft quiet floor for our playground. One day as I was coming up from the brook and brushing my way through the "Pine Forest" fringe, a little yellowish bird with a black mask over its eyes—such a mask as a bandit is supposed to wear—popped out on a White Pine branchlet not two feet from my face. Then it turned and dashed back into the thick foliage again. I had a full view of it in the few seconds that it took to appear and disappear. I didn't know that a bird of such striking pattern could be found any closer than Darkest Africa. It was, of course, the Maryland Yellow-throat of those days—the Northern Yellow-throat as it is now called. I think this must have been the chance meeting that clinched the case and made me a confirmed bird lover thereafter.

I returned to the city in June for my sister's wedding and while in town I bought Chapman's *Handbook*—the bird student's bible of that era—through the pages of which I identified the birds I have mentioned and many others that I met subsequently. Once I had Chapman's *Handbook* as a guide, I made acquaintance rapidly with many birds and learned to call them by their right names. I also acquired a pair of field glasses—an absolute necessity for identifying the smaller birds—and began scouring the Van Cortlandt Park area of New York City, particularly the famous swamp that was a haven for many interesting species. With the *Handbook* and the field glasses, in a few months I was well launched as an enthusiastic observer of birdlife, a pursuit that has made me the butt of amiable mockery now and then, but I have blithely gone my way unheeding for the simple reason that I have had much more fun with the birds than the mockers have had with me.

2

QUINCE
Is all our company here?
BOTTOM
You were best to call them generally, man by man, according to the scrip.

MIDSUMMER-NIGHT'S DREAM, ACT I, SC. 2

It probably is true that a man sees more things and makes more searching observations in the field when he is alone, but there is a virtue in companionship that makes up for any decrease in the supply of clinical notes. A pleasure shared is a pleasure doubled. I always liked to have companions on my tramps through the woods, my walks through the fields or my trips to the seashore. My earliest companions in the pursuit of birds were two of my younger brothers, James and Laurence. My sisters merely looked at birds with kindly curiosity, with a little extra attention to large birds or those that were brightly colored. Sparrows they ignored, though

20

I once met an artist who told me that, if he were rich enough, he would like to retire and spend the rest of his life painting all the sparrows of New England because they were to his taste the most beautiful group of native wild birds. I don't know that I would go that far, but I do think highly of the sparrows. In view of the birdbaths that I kept filled for them and the food supply that I lavished on them through the bleak seasons of many years, I trust that this good opinion is mutual.

I find it amusing, considering my later career, that the first time my name ever appeared in a newspaper it was in connection with a report on local birds. In December, 1913, my brother James and I were standing in front of a cage in the Bird House in the Bronx Zoo and discussing the birds behind the bars. They were native birds, which was why we were so much interested. We were comparing the name plates on the outside of the cage with the birds inside the cage and we found a discrepancy. Among the name plates there was one implying the presence of a Northern Parula Warbler behind the bars. We looked over all the inmates of the cage and could discover no Northern Parula. On the other hand, in our search for that bird we saw a rather soberly dressed little bird that, after a moment's discussion, we identified as a female Black-throated Blue Warbler by the whitish patch at the base of the primaries or outer flight feathers in the wing. Our discussion on this point was overheard by three strangers who were standing close to us and inspecting the same cage.

21

"Hello," said one of the strangers. "You boys seem to know something about birds."

They were middle-aged men of friendly appearance. One of them was short and stout and crippled to the extent that he hobbled along on crutches. We allowed that we knew a bit about birds and told them that if they were interested we could show them about a dozen species of wild waterfowl on Jerome Park Reservoir some miles to the westward. The reservoir was just behind the old homestead in which I was born and I kept a daily watch on the ducks, gulls and other birds that stopped there in passing. The three strangers agreed that they would love to see a White-winged Scoter, a Hooded Merganser, some Redheads and a few more of the species we promised to deliver on the water, so all five of us set out to walk from the Zoo to the Jerome Park Reservoir in the snow. I remember that the stout gentleman on crutches had a hard time of it, but he stuck it out cheerfully and was enthusiastic over the variety of waterfowl he found on a reservoir in New York City.

This lame gentleman—it turned out—was a proof-reader on *The Evening Mail,* a newspaper that has since vanished from the journalistic scene, and there was a writer on the paper who conducted a column titled "The Office Window" on the editorial page. The columnist occasionally had Nature notes scattered through his literary output and our lame acquaintance supplied him with some of these Nature notes from observations he gathered on just such expeditions as he had made with us. For purpose of identification in print, our lame friend was referred to by the columnist as "The Pelham Observer."

In due time "The Pelham Observer" made his report to his friend, the columnist on *The Evening Mail,* and about a week or so after the informal meeting in the Bronx Zoo, I found myself exposed by name in "The Office Window" as a young fellow who knew where interesting species of wild bird were to be found in New York City. I was confoundedly excited and puffed up with pride when I read the article. Others read the article, too, and in a few days such experts as Ludlow Griscom and Charles Rogers from the American Museum of Natural History were peering through their binoculars at the odd assortment of waterfowl on the Jerome Park Reservoir. I think we saw fourteen species there during the Winter of 1913–14. Through Charles Rogers I became a member of the Linnaean Society of New York in 1914 and have been a member ever since.

It was in the Autumn of 1916 that I gave up hunting in favor of enjoying wildlife through a pair of field glasses. I had become a member of the Sports Staff of the New York *Times* in 1915 and a year later—after the World Series of 1916 between the Brooklyn Dodgers and the Boston Red Sox—I went off to Dutchess County on a vacation of two weeks. I stayed at Pat's farm. They often took in boarders and they knew me well. I expected to do some hunting and a great deal of tramping my favorite country in the wonderful Autumn weather. The first day I was afield I carried my 12-gauge shotgun and fired one vain shot at a Ruffed Grouse that got up well ahead of me and made off quickly through the timber. The second day I tramped hill and dale for many hours and never fired a shot. The third day I didn't bother to carry a gun. I just walked for hours and enjoyed the country. Never since that day have I deliberately carried a gun for hunting purposes. Twice I was inveigled into a few shooting incidents, but that was long ago and a story in itself. I didn't give up shooting because I thought it was cruel or heartless. I stopped shooting because I found more pleasure in watching wildlife and enjoying Nature without being cumbered with a gun.

But even while I was quietly enjoying Nature in this fashion, I really knew little about it and was learning very slowly. All my rambles were local. I hadn't made trips to distant haunts of precious species, nor had I attached myself to parties led by registered experts in field observations. It must have been in 1927 or 1928 that, through the fact that we both wrote for the New York *Times* and were enthusiastic about birds, I teamed up with the Dramatic Critic on a December day trip to Long Beach on Long Island. He is a grand companion in the field, the Dramatic Critic; a scholar, a contemplative philosopher with a quiet smile and a sporadic sharp quip that keeps anybody with him ever on the alert. He is a thin chap with a wispy mustache and a fine sense of humor. He laughs at himself most of all, which is the real test of a sense of humor. There is no false front to him, or sawdust stuffing—and he is a distinct authority on Thoreau, wherefor I stood a trifle in awe of him. We walked by the pounding surf in the teeth of a frigid December wind and looked with tearful eyes through shaking binoculars at scoters and ducks offshore beyond the breaker line. We lighted a fire in the shelter of the dunes and heated tea to wash down the sandwiches we had brought with us by way of lunch. We found, in addition to the ducks, scoters, gulls and grebes offshore, some Northern Horned Larks and Snow Buntings in the shriveled herbage of the dunes, and in the clumps of Bayberry we came upon a few lisping Myrtle Warblers.

My only complaint against the Dramatic Critic was that he was not available often enough as a companion on an outdoor jaunt in quest of birds. There were too many new plays coming to the boards. On top of that, he went off to his own country place in June and didn't come back until the Autumn opening of the theatrical season, which would put him on the treadmill again. I reproached him bitterly, but to no avail. He was inconsiderate enough, during World War II, to go off as a war correspondent to China for what seemed an interminable period and, when he returned from that rugged test, he went to Russia. When he was in Chungking I managed to dig up a book on the birds of China and sent it out to him, but when he was in Moscow I could find no book on Russian birds to send him.

The dastardly desertion of the Dramatic Critic was a blow to me, but I was lucky enough to join forces with an old friend and near neighbor, the Artist, who looked the part from the first moment I laid eyes on him, which was many years ago. At that time we were both of a modest age and size, but he was wearing a striking dark beard that made him look—at the very least—like four or five of the Twelve Apostles. The beard is now, as Horatio described that of Hamlet's father, "a sable silver'd," but the Artist still has an eye like a hawk for spotting birds in the field and the tireless foot of a Red Indian on the trail. Another trait that endears him to me is his vast enthusiasm for all sports. He was a fine baseball player in his youthful days in Springfield, Mass., and one of the best skaters in that area. I learned to my cost that he was a wicked opponent in an ice hockey game and I have the scars to prove it. I might add with Sir Andrew Aguecheek: "Though I struck him first, yet it's no matter for that."

One day while I was at my desk in the New York *Times* office, tapping out my sports column for the next morning, the telephone rang and when I answered I heard a booming voice inquire:

"Would you like to see a *Bubo virginianus virginianus* on the nest?"

It was quite a running broad jump from whatever sports topic I was working on to *Bubo virginianus virginianus* in the raw over a telephone wire with an unknown person at the other end, so I stuttered and stammered and finally managed to ask whether or not it was an owl's nest that was in question.

"Right!" said the stranger triumphantly. "Great Horned Owl—about thirty miles out of the city—glad to

26

drive you there on Sunday morning if you care to go."

Of course I cared to go. We made the appointment. That was how I came to meet Herman the Magician, as we call him because his first name is Herman and the Artist and myself, after being with him in the field, decided that his flair for the dramatic was enough to justify us in naming him after a stage wizard who advertised himself as "Hermann the Magician."

Our Herman the Magician holds a responsible position in the Civil Service and has a wife and three children who are his pride and joy. Beyond that, he has four furious pursuits. Owls for one. He dotes on owls, from the little Saw-whet not much larger than a sparrow to the fiercest feathered hunter on the loose in our territory, the Great Horned Owl of our introductory telephone conversation. Most gunners—and "our Owl Friend Herman," as we sometimes called him, does a bit of shooting in the Autumn—hate the sight of a Great Horned Owl and let fly with both barrels as soon as they catch sight of one. But Herman the Magician would no more hurt a Great Horned Owl than he would trample on an orchid, and the wild orchid family is the second of his four furious pursuits.

He is an orchid expert of the most ebullient kind and will rise from close inspection of boggy ground to point downward in triumph and shout exultingly at an aghast companion: "Pogonia ophioglossoides!" He will rave over some native species that doesn't look nearly as pretty to us as the Common Daisy or the New England Aster, but it is sacred to him because it is an orchid, a

27

tribe to which he devotes himself as the Parsees worship the sun.

The third of Herman's four furious pursuits is fishing. He is a fierce fisherman and has all the costumes, trappings, gadgets and weapons that go with it. He loves to catch fish, eat fish and talk about fish. If any careless bystander gives him an opening, he will rattle off the common and scientific names of all the fish, large and small, in adjacent waters, the different tactics used to catch the different species and how each species is best prepared for the table. He is good for three hours any time on trout alone—Brook Trout, Brown Trout, Rainbow Trout, dead or alive—and, unless his listeners flee with horror or collapse from exhaustion, he will go on from there to discuss perch, bass, pickerel, catfish and eels. He talks eloquently and savagely about Snapping Turtles, too, but his vindictiveness toward Snapping Turtles may be set down as professional jealousy. The Snapper is an efficient fisherman on his own hook and gobbles up many trout and bass that otherwise might go into the creel of Herman the Magician or some other devout disciple of Izaak Walton.

The fourth of his furious pursuits is the fern family. From the depths of bogs to the tops of mountains he rambles in search of native ferns in their home haunts. He knows where to find the tall Ostrich Fern and the tiny Maiden-hair Spleenwort. He roams rocky ledges at risk of life and limb to collect uncommon species. To go on a fern-hunting trip with him is an adventure, and to get back home in one undamaged piece is sheer luck.

It was through Herman the Magician that the Astronomer—a real one out of a planetarium—came to be one of our regular and cherished companions on our tramps by hill and dale. I knew his reputation as a scientist. I had sat at his feet when he lectured in the planetarium. At first warning that he was going to join us on a field excursion, I feared that things had taken too serious a turn for us and that we would be crushed by the mere presence of a real scientist in our party—

> But a merrier man,
> Within the limit of becoming mirth,
> I never spent an hour's talk withal.

It was on a cold Winter day that we met the Astronomer. He astonished us at first because he was wearing no hat. He scorns a hat—never wears one—he has a magnificent crop of wavy white hair that serves him as sufficient top covering in any weather. But, at first glance, the absence of a hat in Midwinter was striking. On second glance, the old coat he was wearing was even more striking. It was an ancient corduroy coat of noble birth and he had worn it in rough weather for more than thirty years. He had bought it in Lapland when he was there on a botanizing expedition like Linnaeus—the Astronomer began his scientific career as a licensed botanist with a Ph.D. from Johns Hopkins for a thesis on botany—and it bore honorable scars earned in hard service on long trips through distant lands, including one trip to Siberia to photograph an eclipse of the sun.

There were rips and tears in the coat caused by hard

climbs, sudden falls and occasional chance meetings with barbed-wire fences, and one great gap in the corduroy fairly invited the pointed comment: "See what a rent the envious Casca made!" But for all that, it was a warm coat, a rugged coat, a comfortable and cherished coat, and the Astronomer was determined to cling to it as long as it would cling to him. He held it in the same nostalgic affection that stirred jovial Pierre-Jean de Béranger to write—and please excuse his French—of an old suit with which he could not bear to part:

> *Sois-moi fidèle, O pauvre habit que j'aime!*
> *Ensemble nous devenons vieux.*
> *Depuis dix ans je te brosse moi-même,*
> *Et Socrate n'eût pas fait mieux.*
> *Quand le sort à ta mince étoffe*
> *Livrerait de nouveaux combats,*
> *Imite-moi, résiste en philosophe:*
> *Mon vieil ami, ne nous séparons pas.*

However, I was careless enough to describe the famous coat in my newspaper column, whereupon the Astronomer's wife, who had been casting baleful glances at it for several years, said, "That clinches it!", wrested the tattered garment out of his protesting hands, consigned it to oblivion and made him buy a spanking new coat with all modern improvements.

That completes the tale of our regular strolling party, and week after week for years we walked foursquare to all the winds that blew—the Artist, Herman the Magician, the Astronomer and I. We were fortunate enough to have as occasional additions to our party such men

31

as Edwin Way Teale, Dr. Robert Cushman Murphy, Major George Miksch Sutton, Dr. James P. Chapin and other distinguished naturalists. There was also the young man whom we called the Medical Student, which he was at the time, but it was not long before the Medical Student achieved a degree and went further afield as a medical officer attached to the Army Air Force in the Philippines.

Another part-time companion of ours was an air pilot out of World War I, a member of the famous Lafayette Escadrille. We call him the Falconer because he is devoted to the ancient sport of falconry and has an affectionate interest in hawks, owls, eagles and other birds of prey. He usually has a hawk or two in training or simply as a boarder at his home in the suburbs of New York City.

The Falconer gave us a fine show one October morning. It was a glorious day for a hawk flight and we headed for a high ridge along which the southbound hawks might be expected to glide with the aid of a favoring wind and the "thermals"—rising currents of warmer air—that help the birds on their long journeys. It was a day to bring to mind Kipling's line: "The wild hawk to the wind-swept sky." Just as we turned into the open meadow on top of the ridge we saw an owl on a tall pole. It was a stuffed Great Horned Owl lashed to the top of the pole where it could be seen from a great expanse of sky. Before we could recover from our surprise, we heard somebody say, "Good morning, gentlemen!" and our friend, the Falconer, came out of hiding in a nearby clump of Panicled Dogwood. He had put up the stuffed owl to lure the hawks in for a closer view as they passed in southward flight. An owl, dead or alive, usually will draw the protesting attention of all other birds that catch sight of it. The Falconer reported that, before we arrived, only one migrant had swept in to look the owl in the eye. It was a Cooper's Hawk that—in the veteran airman's language—had flown two tight turns around the owl and then resumed its southbound trip. We concealed ourselves in the dogwood thicket with him and awaited developments.

There was a lake below us and we were watching an Osprey flapping meditatively over the water when we heard a loud whistling cry and turned in time to see a beautiful Red-shouldered Hawk swoop in to toss this defiant note in the face of the impassive owl atop the pole. About twenty minutes later an immature Red-tailed Hawk suddenly appeared from among the trees to the north and made a halfhearted swoop at the owl. The next close visitor was a little Sharp-shinned Hawk that made a couple of impudent passes at the much larger figure on the pole. Then along came the bigger cousin of the Sharp-shinned, a Cooper's Hawk, to investigate this poseur on the pole. The newcomer swooped directly at the stuffed lure with talons extended and, when the target of the menace didn't even blink, the hawk whirled quickly and settled on a tree nearby to stare at the impassive owl for a minute or two, as if trying to make head or tail of such a strange bird. After this inspection, the hawk took off with a contemptuous flirt of its tail and went rapidly down-wind over the valley. We thanked the Falconer for the show. It was our first experience with a stuffed owl as a lure to bring down migrating hawks for closer observation.

The Falconer completes our cast of characters or, at least, the parties of the principal parts, the roster of the regulars of our road company. Others that we encountered or enlisted from time to time might be entered on the program as woodsmen, hunters, fishermen, merry villagers and ladies and gentlemen of the chorus.

3

When icicles hang by the wall,
And Dick the shepherd blows his nail,
And Tom bears logs into the hall,
And milk comes frozen home in pail.
Love's Labour's Lost, Act V, Sc. 2

Full knee-deep lay the Winter snow as the Medical Student and I started out on New Year's Day to make note of what birds we might find in the vicinity. The snow was new and clean and soft underfoot. There would be tracks in it when we reached wild territory. We knocked at the Artist's door but he said reluctantly that he couldn't go along; he had just started on a fresh canvas and he felt it incumbent upon him to stick at the easel. That was his error. Not more than half a mile from his house we found the neat footprints of a fox in the snow. Whether it was a Gray Fox or a Red Fox we had no way of knowing. We have both in our area. But a fox track in New York City—barely a mile from a

35

subway station—is something to gloat over, and in reporting the matter later to the Artist the impulse was to repeat gleefully what Henry of Navarre said to the valiant Crillon, who somehow missed the great battle at Arques: *"Pends-toi, brave Crillon; tu n'y étais pas!"*

There is, of course, no mistaking a fox track for that of a dog, which is a gross, lumbering animal compared to a fox. Reynard leaves his delicate footprints behind him in almost a straight line, whereas a dog leaves a pattern of parallel prints that overlap sloppily and are often askew. A cat comes closer to leaving a track like a fox, but a cat takes shorter steps and would have to brush through snow of some depth over which a taller fox would trot with calm dignity. The Medical Student and I followed the fox trail for a hundred yards or so. We could see that Reynard had been searching for mice, but nothing came of it up to the point where we abandoned the trail and crossed the railroad tracks to walk along the banks of the Hudson River.

We knew we would find some ducks and seagulls on the river. If there were any distinguished visitors among them, we wanted to start the New Year right by "logging" them. I keep a record of the birds I see each year and the date on which I first see them, to which I add in the case of many migrant species that date on which I last see them in the Autumn. It's remarkable how closely the migrants stick to a calendar schedule year after year. Where there are any considerable variations in such dates, the variations are probably my fault; I wasn't on the job to make note of what I should have seen.

36

But we weren't expecting migrants on New Year's Day. We were merely looking around for permanent residents and Winter visitors. Going down through the woods and along the open stretch of brush before reaching the railroad tracks and the riverbank, we had seen some Crows, half a dozen Black-capped Chickadees, a Downy Woodpecker, a Hairy Woodpecker, two White-breasted Nuthatches, a flock of twittering Slate-colored Juncos, about a dozen Goldfinches, a few Blue Jays, two Song Sparrows, five White-throated Sparrows and a cheerful group of Tree Sparrows calling musically as they flew from one tall dried stalk to another where they were feeding in a patch of Great Ragweed. There were, of course, many House or English Sparrows and innumerable Starlings to be seen before we left the residential area for uncultivated territory, and just as we crossed the railroad a Red-tailed Hawk sailed lazily overhead.

The river was dotted with ice floes, some in great packs and some in scattered cakes of all sizes. We swept the ice and the open water with our field glasses for gulls and ducks. When they are not flying, the ducks prefer to stay in the water, whereas the gulls prefer to sit on the ice. It always has amused me to watch ice cakes floating down the river with a dozen dignified gulls getting a free ride. Sometimes we see one gull standing solemnly on a small cake and gazing steadily ahead, looking for all the world like a ferryboat captain in command of his gallant craft.

There were several thousand Herring Gulls on the

ice or in the air over the river. The Herring Gull is the common cold-weather gull in our vicinity, but we seldom fail to find a few Great Black-backed Gulls if we are patient and go over the river slowly with our glasses. These "Black-backs" are easy to pick out among the paler Herring Gulls because they wear, like Hamlet, an "inky cloak" or black mantle, the feature that gives them their common name. They bully the smaller Herring Gulls into surrendering some of the food they find by scavenging. Highway robbery is a popular pastime in the wild, and the Great Black-backed Gull makes a career of piracy over the full extent of its salt-water range. All the "Black-backs" and most of the Herring Gulls leave the river for more northerly waters in warm weather, and the Summer representative of the gull family on the river is the small, handsome and, in season, black-headed Laughing Gull.

When we had taken stock of the gull situation—we had spotted two "Black-backs" among a horde of Herring Gulls on a distant floe—we searched the water for ducks and found just about what we expected. Far out in midstream we saw lines of American Mergansers, gaudy males and sober females, swimming slowly against the tide. Here and there we spotted a Black Duck. Then we found a group of American Golden-eyes feeding by diving. They were appearing and disappearing as though they were being manipulated by a magician. The male American Mergansers and American Golden-eyes are strikingly handsome ducks in Winter plumage, and it was good to see them on the water, but the Medical Student wasn't satisfied. He wanted to see an eagle. He didn't expect to see a Golden Eagle, of course, but he knew that Bald Eagles were regular Winter visitors on our stretch of river. We had walked about a quarter of a mile up the riverbank when the Medical Student, peering through his glasses, said:

"By golly, I think I have one—away over near the far side—on the north end of that big ice floe—see it?"

As I adjusted my glasses I remarked that an eagle on a distant ice cake looks like a chunk of dark timber, an abandoned keg or a lost coal scuttle. This is particularly true of the adult Bald Eagle, whose white head and tail feathers blend with the surrounding ice and snow. When I am looking for eagles on the ice I sweep the floes with my glasses until I see something that looks like a far-off coal scuttle and then——

"You'll have to hurry to see this coal scuttle," interrupted the Medical Student. "It has just taken off and is flying low up the river."

Sure enough, it was a mature Bald Eagle, and the Medical Student, lowering his glasses, said that we had started the New Year in proper fashion by seeing such a great bird. Later in the season the Artist and I saw six Bald Eagles at once along the same stretch of river, four of them sitting on one cake of ice. We soon discovered the reason for this aquiline conference. One of them had a chunk of food and the others were envious. There was a bit of sparring with talons and wings now and then, but the owner clung to the food and finally flew off with what was left of it.

Like the English poets, we have our favorite "Lake District." It's about twenty-five to thirty miles out of town and we often run up there by car on a Sunday in the Winter. There is usually some open water, even in bitter weather, and ducks congregate there in great numbers. One day Herman the Magician was in the lead as we followed a fox track out on the thick ice, when suddenly there was a loud crack from the ice. It startled us for a moment, but it startled the ducks to a much greater extent. In a few minutes the sky was filled with the crisscross and interwoven flights of Black

40

Ducks, American Golden-eyes, Buffleheads, Ring-necked Ducks, Mallards and American Mergansers. Never before had I seen so many ducks in the air at once —there were several thousand of them—north of Chesapeake Bay.

We also made regular trips to the Westchester shore of Long Island Sound and from suitable points of observation we could spot the three scoters that are regularly found offshore there, the White-winged, the American and the Surf Scoter, the White-winged being by far the most common. On the same stretch of water we usually find great numbers of Black Ducks and Scaups, with a scattering of American Golden-eyes, Old Squaws, Canvasbacks, Horned Grebes and Red-breasted Mergansers. The Double-crested Cormorant is an abundant migrant and once, when we dropped in on our good friend Mike Oboiko, he invited us to look through his telescope at an object on the arm of a channel marker at a distance over the water. We all saw the object but didn't have the faintest idea what it was. Smiling with pride, Mike said it was a European Cormorant. Mike is the last word in authority on the birds of that region and we knew that his identification must have been correct, but at that distance I could hardly recognize it as a bird of any kind and a Snow Bunting close at hand would have given me a greater thrill.

I drove the Artist forty-five miles to show him his first Snow Buntings. I knew a parking ground at a Connecticut beach that on lonely Winter days was a feeding area for these Buntings that, to my mind, fit perfectly

Swinburne's phrase "light flocks of untameable birds." There was a bitter wind whipping in over the water as the Artist and I tramped all over the parking plot and we shivered and shook while we made our search. Not a Snow Bunting could we find. We drove off despondently and stopped at a roadside restaurant for a cup of tea. When we had finished the tea and were about to drive home, I suggested that we go back to the beach and have one last look for our birds. The Artist was as willin' as Barkis and back we went. As we stepped out of the car in the parking lot, I heard the friendly "pr-r-r-rt! pr-r-r-rrt!" that Snow Buntings give as they start up in flight. They were all around us on the ground and we had frightened a few into taking wing for safety. The Artist was well satisfied with the trip.

One day we learned that Snowy Owls had been seen near Jones Beach and we decided to run down and have a look. It's about sixty miles from our headquarters, but it is practically all parkway, so we made it in good time. Edwin Way Teale met us on Long Island and told us he had seen two Snowy Owls on the dunes a few days earlier. As we were rolling slowly along the dune road in our cars, scanning the sand dunes and the patches of dried grass and wind-whipped shrubbery for anything that looked like a misplaced white pillow or a forgotten bundle of wet wash, the Artist yelled for us to stop the car and jump out. When we obeyed orders, he pointed to a bird on the telegraph wire overhead. It was a Northern Shrike, and just as we fixed our glasses on it, the bird left hastily in pursuit of a small bird that we did not identify. We climbed back into the car again and, as we resumed our patrol, we saw hundreds of Pine Siskins feeding on weed seeds close to the roadside.

We had covered four or five miles when we noted that Edwin Way Teale had jumped out of his car up ahead and was flagging us down. When we joined him, he pointed to the edge of the marsh on the bay or north side of the road. We saw a white patch and, with our glasses, identified it as a Snowy Owl staring back at us. We walked toward it, but before we came close, it took off and we saw that it was carrying some prey. It alighted about a quarter of a mile down the marsh on a chunk of timber. The Artist stalked it, Indian fashion, and let it feed a bit before he came up close enough to flush it. When he came back to us, he had some Rabbit hair that he had recovered where the owl had been feeding and he said the bird was carrying about half a Rabbit as it flew away. A few minutes later we saw another Snowy Owl perched on a post in the marsh, but there was no way of getting at all close to it. We walked over different parts of the beach and the dunes in the hopes of seeing Snow Buntings, but the best we could do was to put up some flocks of Northern Horned Larks that let us come within a few yards of them before they took wing.

The Astronomer had his movie camera with him on this trip and, with a telephoto lens, he managed to get some footage of the Snowy Owl whose dinner we disturbed, but it was just three months later—our owl expedition was on December 9 and our next trip down was on March 10—that he made his big haul in pictures. Edwin Way Teale had sent word that he had a fine flock of Evening Grosbeaks coming regularly to a dinner table that he had set up for them behind the public library in his town. He kept the prized visitors happy with a diet of sunflower seeds. I've forgotten how many bushels of sunflower seeds he lavished on the gaudy Evening Grosbeaks before they left for parts unknown, but it was no small amount.

We drove down of a crisp March morning and the birds were there as advertised by Brother Teale, who was waiting to greet us. If the birds had not been visible, perhaps Brother Teale would not have been in sight, either. But there he was on the library lawn with a wide smile and some sixty Evening Grosbeaks spread over the feeding tray and a couple of nearby trees. It was the first time the Artist had seen this species and he was enthusiastic over the males in their colorful costumes of black, yellow and white. The Astronomer had his color film in his camera and, with clear sunlight to help him, he took pictures of the birds from all angles and close range—the birds were quite tame—while Edwin Way Teale told us that he knew where there were some Green-winged Teal, some Shovellers and some other interesting ducks on various ponds not far away.

When we heard about the Shovellers, we couldn't even wait for the Astronomer to finish his color photography. The Artist and I never had seen Shovellers in

the wild and we were in a hurry, so off we went with
Herman the Magician to what had been in the old days
the "ice pond" of the town, and there we found nine
Shovellers, the exact number that Mr. Teale had promised to deliver on the water. While we were enjoying a
view of the Shovellers, two strangers with field glasses
came along and told us that they had peered through
a wire fence at a private pond a few miles off and had
seen Gadwalls among the other ducks there. Gadwalls!
Another new species for the Artist and me. We waited
for the Astronomer and Brother Teale to catch up with
us and off we went for the Gadwalls. It was such a
private pond that we had to shoulder our way through
a privet hedge before we could get to the wire fence
that kept intruders from approaching the water. We
had been told that the owner put out food for the ducks
and that it was a haven for migrants and Winter visitors
of the duck family. There must have been several hundred ducks on the water and the bank of the pond. They
were mostly Black Ducks, but we soon picked out half
a dozen Gadwalls—grayish ducks with dark tail areas.
They have white speculums and a touch of chestnut
along the wing, but this we did not notice. The easiest
"field mark" was the dark area near the base of the tail,
and I was surprised to find later that this was not fea-
tured by Forbush in his great work on New England
birds.

After we had looked over the Gadwalls we shifted
our glasses all over the place and I found four small
ducks asleep on the bank with their heads tucked under

their wings. They were the right size for teals, but they couldn't be Green-winged Teal or a white "shoulder mark" would have shown. They had to be Blue-winged Teal if they were any native duck that we knew. On close scrutiny, I detected part of the white crescent—just the tip of it, barely enough to notice—on the head of one bird, and that was enough to identify it as a Blue-winged Teal. Then, to make it easy, something roused the birds and they marched down to slip into the water. Even so, they did not show in walking or in swimming the light blue shoulder patch that is so conspicuous when they are on the wing. One day in Van Cortlandt Park I was scanning the reeds when four ducks shot across my field of vision. I saw them for only a second as they flashed by, but the light blue of the wings "hit me in the eye" in that time.

We thought we had done well with the Gadwalls and the Blue-winged Teals, but our Long Island guide said he had other sights for us to see if we would follow him, which we did joyously. We drove a few miles to another pond, and there among the ordinary Baldpates we saw one with a shining cinnamon head, indicating that it was a European Baldpate or Widgeon. There were also two Mute Swans on that pond, but this species is half-domesticated (or semi-wild, if you prefer) along all that section of Long Island. They were imported originally as decorative waterfowl for parks and private estates, and the descendants in great number have gone off to roam at will just as so many

European flowers, imported for gardens, "escaped" to enrich our catalogue of native flora.

"Just one more pond to visit," said Brother Teale. "I can't promise to deliver this bird, but it's worth looking for and it was there yesterday. It's a European Teal —if it's there!"

He went off in his car and we followed to a park lake just a few hundred yards from a main motor highway. There were hundreds of Black Ducks, Mallards, Scaups and Baldpates on the little lake. Our guide, who, like the rest of us, was studying the scene through glasses, cried that he had the European Teal in view. I swung my glasses over and found a pair of teals, but the male had a vertical streak of white forward of the bend of the wing—or so it seemed to me—and that would make it our native Green-winged Teal. My eyes are none too good, so I moved closer to the bank of the lake for a better view. As I did so, I noticed another pair of teals much closer and moving away as I walked down to the water. I put my glasses on them and immediately saw on the male, in the lead, the horizontal white streak above the upper edge of the folded wing that distinguished the European Teal. I took it for granted that the female with it was the same species, but it was pure supposition on my part. I never would be able to identify the females, but I presumed that they, like sensible females, knew their place and would stick to their legal spouses when traveling as well as when at home. Just as I called out that I had finally found the European Teal, our Long Island guide remarked that he had found a pair of Green-winged Teal further out on the water. That cleared up the temporary confusion. He had first sighted one pair of teals and I had first sighted the other. It was a satisfactory solution to a mild mystery of the moment and it completed a great day for us, because we don't too often see teals of any kind. We call it a good day when we see one species. On this occasion we had seen four species in one day—the Blue-winged Teal, the Green-winged Teal, the European Teal and the Edwin Way Teale, the last being a noble specimen, too.

These trips to Long Island were special expeditions. The regular route for our strolling band lay closer to home, especially when snow covered the ground and traveling was treacherous. Sometimes the roads were so slippery with packed snow or ice that we stuck to footwork and rambled the nearby terrain, the Van Cortlandt swamp, the wooded ridge to the westward and the wind-swept banks of the Hudson River. It was bitter along the riverbank when the wind was sweeping in from the northwest and tears would stream down our cheeks as we held our glasses to our eyes with freezing fingers and looked over the ice floes and open water for ducks and gulls. But when the roads were reasonably safe for cars, Herman the Magician would drive us up to the "Lake District" in Westchester and we would inspect the concentration of ducks in what open water there was around the inlets, or tramp the woods and fields just for the sake of the wintry scenery or the fun of being out in the open. Any birds that we came upon were just that much added fun.

For we always did have fun at it, no matter what the weather. For years we had been going out in all sorts of weather, and probably many respectable persons who stared at us out of passing cars or from the warm windows of happy homes thought that we must have been slightly touched in the head, a little on the weak side mentally, to be wandering about in what looked like fearful weather. But we long ago discovered that bad weather always looks much worse through a window. We were dressed for it and, except that we never gave three cheers for bitter cold or a driving rain, we could take it all in stride. We loved to be out when the snow was falling. A woodland covered with a fresh fall of snow, when each tree and bush—limb, branch and twig—"wore ermine too dear for an earl," was to our eyes a glimpse into the fairyland of our childhood dreams. And all the while on these trips we bantered one another on matters personal or political, on art and athletics, on Shakespeare and the musical glasses, and the Astronomer—who might have crushed us with the weight of his scientific knowledge—was always the jolliest of the lot.

4

BRUTUS
Is not tomorrow, boy, the Ides of March?
LUCIUS
I know not, sir.
BRUTUS
Look in the calendar and bring me word.
JULIUS CAESAR, ACT II, SC. 1

Somewhere in his writings Thoreau set it down that the human ear is not keen enough to catch the first footfalls of returning Spring. Perhaps so. Or perhaps Spring never quite goes away and is always with us to some slight degree. Who knows the first instant when the buds begin to swell on the trees or the very moment when the Skunk Cabbage first wriggles its hooked nose and starts to push up through the debris of the frozen ground?

I like to think that Spring begins on the tick of the Winter Solstice, when the sun "starts north" again. It isn't until the middle of January that there is any great difference in the height of the sun in the sky at noon or any noticeable earlier rising or later setting of the sun. But definitely—and astronomically—the Winter Solstice marks the turn for the better, if we look at it in that light despite the fact that practically all the hard weather of Winter lies ahead of us at that time. That, however, is taking a rather lofty view of things. Underfoot, conditions are more realistic. We begin to look for signs of Spring about Lincoln's Birthday or Washington's Birthday and, aside from the swelling of the buds and the Skunk Cabbage tips poking up in marshy ground, one of the early signs is the nesting of the Great Horned Owl.

One morning in late February we were walking a wooded ridge and peering at the ground around the biggest Hemlocks in the hope of finding owl pellets that might indicate a nest in the vicinity. We knew that these great Hemlocks housed owls throughout the year, but it was a difficult matter to catch sight of them in any season. We had just turned to go down a slope when the Artist said: "Whoa! What's that up yonder?"

We looked overhead where he was pointing and saw a mass of leaves in a Hemlock crotch, with something like a short blunt object sticking over the edge.

"Might be an owl's tail!" said Herman the Magician, who is the leader on all our owl hunts. "Somebody skin up that big rock and take a look."

The Medical Student was along. He clambered up the rock formation to a height that put him on a level with the mass of leaves in the tree and swung his field glasses on the target. A moment later he yelled excitedly: "Holy Moses! It's a Great Horned Owl and it's looking right at me!"

We all scaled the big rock and had a look. It was February and there was snow all over the ridge. The temperature was well below freezing. Yet the Great Horned Owl already had set up housekeeping, truly a hardy bird. It might as well be recorded here that the hatching came off successfully and we saw the two young birds grow up to full size on a plentiful diet of—from the evidence we found on the ground—Rabbit and Pheasant. Shortly after they left the nest one of the young owls was found dead on the ground without a mark on it, and we never could learn the reason for its untimely end.

We tramped the ridge regularly in March for the purpose of seeing how the owl was faring on the nest. On the way we passed through a swampy patch and along a sheltered hillside on which we found the advance guard of the host of Spring birds and flowers. On any warm March day we were sure to hear the Song Sparrow and the male Red-winged Blackbird. The Phoebe was hovering along the fringe of the woods and the Golden-crowned Kinglets were busy in the Hemlocks on the ridge. We found Robins along the roadsides and on lawns, and a few Meadowlarks sprang up from the dried grass on flat patches of ground. Dandelions and Coltsfoot were blooming in nooks that faced southward and, in the last week of March, the undergrowth at the head of the pond that lies below the ridge was golden with the flowers of the Spice-bush. On March 30, which was warm and sunny, we found the Rue Anemone, Bloodroot—what a revolting name for such a delicately beautiful flower!—and Hepatica in bloom. That same day we saw a Mourning-cloak butterfly flitting past us in the woods and we came upon two Mourning Doves on a path picking up little bits of stone, presumably as an aid to digestion. Bluebirds were on the wing and all morning their plaintive warblings came drifting down from the sky. There were Field Sparrows and Flickers about and the White-breasted Nuthatches were keeping company with flocks of Black-capped Chickadees among the Hemlocks on the ridge.

The Red Maples, of course, were draped with their flowers and we looked at them through our magnifying glasses. There is nothing, I think, that pays richer dividends for a longer time than the investment of twenty-five cents in a magnifying glass of a few diameters. We turned them on flowers and insects and it was astonishing how some of these objects looked under the glass. It's only with the aid of a magnifying glass that the marvelous structure of the pistils and stamens and other parts of some of our common flowers can be appreciated. Just try it sometime. Satisfaction is guaranteed.

The first sign of rich green on the ground in our area is the sprouting of the Wild Garlic in the meadows and on the floor of the wet woods. I avoid it as a matter of personal taste—or distaste—but Herman the Magician pounces on it and gathers up handfuls to be carried home to flavor his soup. He does the same with the broad-leaved Wild Leeks that appear in lush patches about the time that the Shad-bush breaks into bloom. The Shad-bush is so called, of course, because it is in flower about the time that the Shad run up the rivers to spawn.

On the last Sunday in March we were rambling up the west bank of the Hudson River at the foot of the Palisades when we encountered one of the Shad fishermen who come there yearly to string their gill nets in the river and take the fish to market. The fisherman we met was a grizzled veteran who was smoking a pipe and taking his ease ashore just a few paces from the flatboat on which he lived and which was tied alongshore. He said he was from the Jersey coast—that the Shad run in the Hudson usually lasted about two months—that when it was over he would go back to offshore fishing off the Jersey coast—that a nine-pound fish would be a nice one for his net—that there was a good market for fish at the moment—that he was in no hurry to go to work that morning because he had to wait for the tide, anyway, and nobody could hurry the tide.

The Artist said that fish were just as queer as humans. Here were the Shad coming up the river to spawn each spring and then there were the eels that lived in fresh water and went downstream to the ocean to lay their eggs in the Sargasso Sea, a fatal journey from which they themselves never returned. I said that I

remembered reading in Forbush that the Laughing
Gulls return to New England in the Spring at about the
time when the Alewives run up the rivers.

"What is an Alewife?" asked the Artist.

There he had me. All I knew was that it was a fish
of some kind, but I never had been properly introduced
to one. When I reached home after our visit with the
fisherman I picked up my little cloth-covered handbook,
Representative North American Fresh-Water Fishes,
by my friend John T. Nichols—"as tall a man as any's
in Illyria"—and looked through the index without finding any trace of "Alewife." I looked in a dictionary and
found "Alewife" defined as "a species of small fish of the
anadromous type." Then I thumbed the dictionary to
discover what "anadromous" meant. I found it without
trouble. "Anadromous: Running up, ascending; said
of fishes, as salmon, that go from the sea up rivers to
spawn." Well, in that case, the Shad would be in the
same category. I turned back to the Nichols handbook
and confirmed my suspicions by reading on Page 27
that the Shad (*Alosa sapidissima*) was an "anadromous"
fish.

By this time I wanted to get at the bottom of this
business of running up or down rivers to spawn, so I
traced the down-running eel to Page 29 of the Nichols
booklet and found the slithery eel—a true fish, of course
—stigmatized as a "catadromous" fish. I fumbled around
with the dictionary again and traced it all back to the
Greek "dramein," meaning "run," with the prefixes
"ana" and "cata" originally "kata"—to indicate which

56

were the up-runners and which were the down-runners. Thus the chance encounter with a man who hauled up Shad from the river drove me to the unrelated task of digging up Greek roots.

About the last of March the Artist said it wasn't too late to tap a few Sugar Maples and boil out a cake or two of maple sugar. Against this I set my foot firmly. Like the ghost of Hamlet's father discussing another unpleasant matter, on the subject of maple sugar I could a tale unfold whose lightest word would harrow up the Artist's soul and cause his hair to stand on end like quills upon the fretful porpentine, more or less.

"Go ahead; don't mind me," said the Artist.

So I told him how I had once made a mess of maple sugar during the period when I was teaching school in Dutchess County. That's a hardwood sector and the sap run of the Sugar Maples is quite an event. Some farmers make a real business of reducing the sap to maple syrup or maple sugar. Every farmer has at least a sap pot on the back of the stove to make syrup or sugar for home use. We had hundreds of fine Sugar Maples on our farm—a dozen in our very dooryard—and my brother and I decided that we would make some maple sugar.

We went down to the general store and bought the "spiles" to drive into the trees and we gathered up a wild assortment of pots, pans and pails to catch the sap—milk pails, water pails, lard cans, pots with swinging handles, catch basins of any kind. For our big boiling operation we took the galvanized iron washtub, family size—

57

and we had a large family. We decided to do the boiling down in the wood lot, right in the thick of the source of the raw material. We set the "spiles" and the pendent pails in position on a windy morning and went off to school. Late in the afternoon we were back to gather the sap and pour it into the big boiler, which we had set up on stone piers with room underneath for a wood fire. The stone piers were roughly constructed and the boiler, when we poured in the sap, wobbled dangerously but not fatally. We set the buckets back in place to catch more sap, lighted a fire under what we had in the boiler, stoked it to last as long as possible, and went home to supper. We were back in the woods an hour later, by which time it was pitch dark around us. We had blankets and a kerosene lantern. We were going to stay there all night and keep the fire going as we went from tree to tree and gathered more sap for the boiling. We could take turns working and sleeping.

The first thing we discovered as we stumbled around in the dim light of a smoky lantern while "the wind was a torrent of darkness among the gusty trees" was that we should have had covers on our sap buckets, and on the big tub in which we were boiling the sap. Whipped in the wind, the branches of the trees were wailing and groaning and shedding bark and twigs in all directions, including into our sap supply. Insects of all kinds managed to find their way into the sap too. We had no strainer of any kind. Whatever foreign matter went into the sap went into the boiling.

It was cold in the windy woods. About midnight it

began to rain steadily. Stumbling in the rainy dark, I fell and smashed my wrist watch. My brother mislaid one of his mittens, which handicapped him at the cold work of carrying the sap buckets to the boiler and back to the trees again. We had no shelter from the rain among the leafless trees and soon we were soaked to the skin, but we managed to keep the fire going and we could see dimly by the light of our lantern that we really were boiling down the sap.

Came the dawn—a cold slow dawn with the rain still falling relentlessly. The washtub contents—replenished so often under difficulties in the darkness—had simmered down to an ugly-looking mess of negligible dimensions, hardly enough to cover the bottom of the boiler. We tilted it to concentrate the residue into a "corner" of the tub and then left it to cool while we gathered up our paraphernalia. When we came to pour the "syrup," the confounded stuff wouldn't budge. It had set as hard as reinforced concrete or native New England granite. We finally belted it loose with some wild wallops from the ax, damaging the washtub no little in the process, but by that time we were desperate. When we had the chunk in the open, we tried to taste it but couldn't dent it with our teeth. We cracked it open with the ax and in the center we found my brother's missing mitten. We also found embedded in the body of the concoction a notable collection of bark segments and sugar-coated beetles. We heaved the stuff away and retired abruptly from the maple sugar trade.

The Artist said that we were very crude operators and had horribly botched one of the great native arts of New England, the production of the most luscious delicacy in the world, but he bore no malice and would be glad to talk about the weather to ease the strain. He said that probably it was the March weather that made the March Hare mad. One day there wasn't a breath stirring and the next day it was blowing half a hurricane. Some March days are warm and soft and sunny, and others are cold and hard and soaked by a dreary rainfall. But Spring comes creeping onward persistently, despite the zigzag pattern set by the Weather Man. The Pussy Willows burst into velvety bloom by the calendar and the *cro-o-onnk, cro-o-onnk, cro-o-onnk* of Canada Geese going northward by day is echoed at night by the piping of the frogs in the marshes.

One March day we went over to the Van Cortlandt Park lake and, from the west bank, saw ducks scattered all over the water. The sun was in our eyes, which put the ducks in a bad light for identification. We could see well enough, however, to recognize Scaups, Black Ducks, Mallards, a few Pintails and some Baldpates. Among the Baldpates we thought we saw one with the reddish-cinnamon head that would make it a European Widgeon. We had to investigate further to make sure. We decided to go around the head of the lake and have another look at the doubtful duck with the sun in our favor. We made the loop around the upper end of the lake through a clump of willows and, as we came out in the clear, we saw a few Canada Geese in the water quite

60

close to us. We paid no attention to them because we were after other game. We were skulking along the east bank in an attempt to reach the most favorable position for a good look at the doubtful duck—it was really a drake, to be sure—when suddenly there was an unholy hubbub of clatters and squawks nearby. We turned and saw that two of the Canada Geese—rival ganders, we took them to be—were having a fierce fight. Each had fastened his beak on the other bird's head feathers and, from that position, they were pulling and hauling and squawking in a most disorderly manner. Then, coming breast to breast, they rose half out of the water and began to beat one another with their powerful wings. There was much sound and fury and feathers flew like dead leaves in the Autumn wind. Finally one of them decided that he had all that he cared to take of that treatment. He pulled out and dived under the surface for safety, with the winner in bitter pursuit. The loser, after some further heckling underwater, escaped to shore, where he pulled himself up breathlessly and bedded down on the dried grass to regain his poise and nurse his wounds. The winner paddled off, honking triumphantly to celebrate his victory. But we suffered with the losing gander because the battle frightened all the ducks on the lake and every single one of them flew off as the first loud blows were struck by the confounded ganders. We did get a look at the doubtful duck as it went overhead in the sunlight and we were fairly sure that it was a European Widgeon, but not sure enough to satisfy ourselves completely. We still nursed a grudge against those ganders.

Thus we walked the windy ways of March that seemed like June one day and January the next, and the wind vexed us, but we knew that without it there might be no life on this planet, so we bowed our heads and struggled forward. Some birds already were back with us, the buds were opening, Crocuses were dotting suburban lawns and in sheltered gardens we saw the golden Daffodils

That come before the swallow dares, and take
The winds of March with beauty.

62

But there is no blinking the fact that it is a rugged month with us.

As Laurence Sterne wrote of another topic: "They order this matter better in France." I much prefer the European March to the North American March, just as I prefer the North American October to any month in any other land that I have ever visited. I spent two years in France and can commend Spring on the Continent. Better still, I once went to England in late March to watch the running of the Grand National Steeplechase at Aintree—it was the year of Reynoldstown's first victory—and the setting was just what the English poets had led me to expect. There was the fresh green of the young grass all around; there were flowers along the banks of the brook that cuts across the famous steeplechase course; the sun was bright; the weather was mild; there were Skylarks rising from the infield to soar and sing in the blue sky overhead—truly this was Spring in the country of gentle Gilbert White, the England of Shakespeare and Tennyson and Masefield. I loved it.

5

---

*The pleasant'st angling is to see the fish
Cut with her golden oars the silver stream,
And greedily devour the treacherous bait.*
MUCH ADO ABOUT NOTHING, ACT III, SC. 1

As March goes out and April comes in, we are sometimes inveigled into dipping for Smelt, a fearful and
wonderful pursuit as we follow it. Our leader on such
trips is Herman the Magician, an expert on all things
fishy. He is our Master of Smelt Hounds and his official
uniform as M.S.H. is a sight to behold. He wears hip
boots, corduroy pants, a thick wool shirt (red and black
checks preferred), three to five sweaters, a hunting coat
lined with sheepskin and a peaked hunting cap. His
ordinary armament consists of a twelve-quart pail hung
on his belt, a dip net with a four-foot handle, and a
flashlight. We once took an innocent veterinarian out
at night on one of these trips and his subsequent report
of the outrage consisted of a single sentence, to wit:

64

"We started off at dusk in the rain and after that everything went black!"

There is a certain amount of mystery connected with the dark pursuit of *Osmerus mordax* (ancient Italian slang for Smelt) in our territory. It appears from the testimony of respected citizens that, in some other regions, there are Smelt that have the decency to "run" or go on their spawning journeys by daylight and also may be taken with hook and line, but in our neck of the woods the confounded Smelt insist on keeping it all as dark as possible; they run only at night and it's illegal to take them with hooks or lines. Irked by this difference in habits, I invaded the American Museum of Natural History, tracked down my tall friend John T. Nichols, Curator of Recent Fishes, and asked him why Smelt in some areas make their spawning journeys in broad daylight, whereas in our territory they run darkly through the gloom of night. Mr. Nichols, who is a Harvard alumnus, had a ready answer. He said: "I don't know."

It's about forty miles from our neighborhood to the rocky stream in the woods where we do our dipping and it's usually a frigid ride in either rain or snow or both. Our valiant Master of Smelt Hounds has his spies who live along the distant stream. When they phone that the Smelt are running, he dons his hip boots, pulls on his heavy "smelting coat," grabs his strongest net, gives the fishy equivalent of "Yoicks!" and comes clattering around in his car to gather up his posse of Smelt pursuers. With the journey to and from the scene of operations, the hunt in the dark actually lasts six or seven hours.

One night in the first week of April the M.S.H. tootled around in his station wagon at dusk. He had the Artist —an accomplished smelter—with him, and the usual assortment of pails, nets, flashlights, rubber boots and blankets to keep all hands from freezing to death on the journey. The ride was cold and dark. After an hour and a half of rattling over the roads, we reached the woods and pulled up alongside a fence. Sitting on a soggy bank, we took off our shoes, pulled on hip boots, hitched the twelve-quart pails to our belts, stood up, brandished our nets, made sure that our flashlights were in working order, pulled down our caps, buttoned up our heavy coats, pulled on our gloves and, thus lightly accoutered, climbed over a wire fence and picked a precarious way down through the woods to the rushing stream up which the Smelt were reported to be running. There was snow in the woods and we had more than one fall from uncertain footing on this part of the journey.

Once when we made this trip we saw the rushing stream dotted with flashing lights as we approached it through the trees. That was proof that the Smelt were running briskly and the whole area was aware of it, the result being that men, women and beasts were out with pails and nets and flashlights and lanterns and this rushing little river in the woods was almost like Broadway with its countless flickering lights. Even I, a most reluctant and inexpert smelter, dredged up about ten quarts of Smelt that night and, with blind sweeps (my flashlight had gone dead), came up with as many as seven Smelt with one dip in the dark. But this later April night there were no signs of life along the stream, no flashing lights to make it a pretty picture. There was merely the increasing roar of the tumbling waters as we stumbled, staggered and slipped downhill through the woods toward the bank of the stream. The lack of light along the stream indicated that no other Smelt hunters were on the job. It also raised the dark suspicion that perhaps the Smelt were not on the job, either. But our M.S.H. is never easily discouraged. He hit the bank of the rushing stream in full cry, sounded his official "Tantivy" and staggered out into the stream with his net at the alert and his flashlight beaming down into the swirling waters to spot his anticipated prey. In no time at all his flashlight was intermittently on and off like a will-o'-the-wisp all over the stream. I cautiously explored a few of the quieter holes and the Artist slowly inspected the slower water along the bank. Nobody sighted a fish for the first half-hour, at the end of which time rain began to fall drearily. The wind became louder and colder and the branches were thrashing wildly over our heads in the thick darkness.

The situation was more than passably discouraging and even the M.S.H., from a dangerous stance in midstream, yelled over the noise of the waters that the Smelt didn't seem to be running. I grumbled gloomily to the Artist that they weren't even walking. But just when things were at their blackest, the M.S.H. raised his "View halloo." He had sighted a Smelt and was in full chase. A few minutes later the Artist stalked and captured a small Smelt in a little pool near the bank. So there were some Smelt in the stream, but the run was dolefully slim. There were no such swarms of succulent Smelt as we had dipped up—a full breakfast order with one swoop—on happier occasions. This pursuit of scattered Smelt was a disappointing and exhausting procedure. A grown man stalking a single Smelt seemed undignified. With this in mind, I raised a loud motion to adjourn. At that precise moment there was a splash in midstream and a light went out. The M.S.H. had lost his footing and gone down in the tumbling waters.

"We're in an awful hole if he doesn't come up," shouted the Artist. "He has the keys to the car in his pocket."

However, our waterlogged M.S.H. made shore in a minute or so and apparently was none the worse for his "header" or "sitter"—he refused to tell which way he had gone down in the darkness—in the icy water. Even such a spill failed to dampen his enthusiasm for the pursuit of *Osmerus mordax*. He even cooked up a scheme— and the Medical Student and I fell for it—to dip for Smelt up to midnight on the eve of opening day of the trout season, sleep out in the woods in his station wagon and rise at dawn to thrash the same stream for that tastiest of game fish, the native Brook Trout—may its speckled tribe increase!

We did dredge up a few Smelt that night, but it was cruelly cold and, though we wrapped innumerable blankets around us when we crept into the body of the station wagon to sleep, I thought we would freeze to death before morning. I lay there shivering and yearning for daybreak so that there would be a good excuse for getting up and running around to restore circulation. We were up at the first streaks of dawn and, while I was pulling on my shoes, I heard the croaking note of some bird in the woods close at hand, a harsh note that I did not recognize. I grabbed my field glasses and went on a hunt in the dim light. Just as I caught the bird in my field glasses, it began to sing—and I nearly dropped my field glasses in astonishment. At the first few notes of its rising roulade, I knew that it was a Hermit Thrush!

Though I had known the bird for many years as a migrant in our neighborhood, this was my introduction to its vocal powers. It does not sing commonly on migration and I had not been observant enough to link the migrating Hermit Thrushes, which are abundant in our area in Spring and Fall, with this discordant call note. But since that cold dawn in the woods when a great light came upon me, I have come to know the song and call notes of the Hermit Thrush very well, because it nests regularly and sings for hours at a time around the shack I built in the Berkshires. And since I have learned its song and call notes, I hear its croaking call note frequently as it migrates through our neighborhood and I even catch occasional snatches of its lovely liquid song, especially in Autumn when it gives a half-whispered version like an echo of Summer.

The main purpose of spending that night in the woods, however, was—as Herman the Magician resolutely reminded us—to thrash the nearby stream for that succulent dainty, the *Salvelinus fontinalis,* or Brook

69

Trout. After the Hermit Thrush had moved on, we readied our rods and waded in. We spent hours stumbling and staggering amid the rocks and the rushing water and never caught a thing. We lashed pools to a lather and didn't even get a rise. The Medical Student and Herman enjoyed a few moments of hilarity when I slipped on a moss-covered boulder and fell backward into the icy stream, but otherwise the trout trip was a complete loss and the whole morning would have been wasted if it hadn't been for the chance meeting with the Hermit Thrush and the wonderful experience of hearing its lovely song for the first time.

I might as well break down and confess that I am not an expert fisherman, nor even an ardent one. I have fished in the Gulf Stream off the Florida keys and three times had Sailfish on my line, but in each case the attachment was not of a serious or prolonged nature. I once wrestled a six-foot Hammerhead Shark to boatside in forty-five minutes off Miami Beach at the urging of the owner of a neat cabin cruiser who was confoundedly sorry he ever mentioned it, because the captain lashed the defunct shark to the side of the boat while we trolled for Sailfish for an hour or more, during which period the rough skin of the defunct shark chafed all the paint from a big section of the hull. The owner of the boat gave a great groan as this was called to his attention and he swiftly cut loose the carcass of the Hammerhead.

In the Gulf of Mexico I had hauled in dozens of Spanish Mackerel and, at odd times, a fair assortment of Kingfish, Amberjack, Sea Trout and other edible fish

70

of those teeming waters, but I did this merely to be obliging and go along with friends. I never stirred up a trip of that kind under my own steam. I like to eat Brook Trout, but I am not one to make a sacred ceremony or a ritual of the catching of them. Since it is too late for the game warden to catch up with me, I will make a horrible confession. In *Twelfth Night*—one of my favorite Shakespearean plays—Maria says of Malvolio: "Here comes the trout that must be caught with tickling." There was an ancient belief that trout could be caught by tickling their bellies as a preliminary to grabbing them firmly with both hands and lifting them triumphantly toward the cookstove. I never deliberately tickled any fish but—*horresco referens!*—I must admit that when I was a boy I caught dozens of Brook Trout with my bare hands. There was a huge flat-topped boulder in a good pool and the current swept along the flat side that ran down into the water. There the speckled beauties used to lie, heads upstream, waiting for food to come to them. I would creep out on that boulder and, lying flat upon it, slowly lower my hands into the water to pin a fish against the flat side of the rock. It required infinite patience, because all the fish would be off in a flash at the first sign of quick movement on my part. But if I moved my hands very slowly, my chosen victim would stay in place until I herded it against the face of the rock, where I could pin it for capture. Even then some of them managed to wiggle free because Brook Trout, being practically free of perceptible scales, are hard to hold with bare hands. Herman the Magician, our piscatorial authority, insists that they do have scales, which I admit. I merely claim that they are not worth mentioning. In fact, one of the good qualities of the native Brook Trout in the eye of an amateur chef is that it does not need to be scaled before cooking.

A bad feature of fishing in the woods is the insect life that accompanies it. The flies are terrific, and assassinating one fly seems merely to draw four or five more flies to the scene of the death. Later in the season hordes of mosquitoes get in their nasty work. I become lumpy on hands and face in no time at all. It's tough when the insects are biting and the fish are not. My only compensation lies in the birdlife along the stream. I carry my field glasses slung around my neck, which handicaps me as I poke my trout rod through thickets or balance myself on logs for a crossing of the stream, but I can drop the rod and swing the glasses into action at need, which is the main point. If I see a few good birds of a morning along the stream, anybody else is welcome to the fish.

6

When daffodils begin to peer,
—With heigh! the doxy over the dale—
Why, then comes in the sweet o' the year;
For the red blood reigns in the Winter's pale.
                              WINTER'S TALE, ACT IV, Sc. 3

We never really discover anything in our outdoor rambles. The best we can do is to uncover a few things new to us but old to wiser folk. "Discovery" is a word of doubtful meaning, even in geography. It is alleged that Columbus "discovered" America, yet there were millions of Red Indians living in this country before he set foot on San Salvador or Watlings Island. I think a delightful definition of "discovery" as commonly used is that offered by the great explorer, Vilhjalmur Stefansson, who said with a devilishly impassive face and a heavenly twinkle in his eye: "Discovery occurs when any land is first visited by a white man—preferably an Englishman."

But our group is not adventurous. We are humdrum and hidebound in our land voyages. We hunt old trails. The Astronomer, who knew John Burroughs, quoted that bearded authority as saying to him: "To find new things, the path to take today is the path you took yesterday." We follow that rule with some slight modification to suit the changing seasons. We know that life begins at the water table and in February and March we scan the marshes, ponds, lakes and watercourses for early birds and flowers, for this is the season when

> *. . . time remembered is grief forgotten,*
> *And frosts are slain and flowers begotten,*
> *And in green underwood and cover*
> *Blossom by blossom the Spring begins.*

The fields, roadsides and thickets are alive with sparrows of all kinds by the end of March. The Killdeer run about the flat ground, piping plaintively. Red-winged Blackbirds call incessantly over the marshes and from the topmost twig of a White Ash a male Cowbird puffs out its throat, bows awkwardly and gives off the ridiculous song that always reminds me of the English dish of cabbage and beef that they call "bubble and squeak." We catch glimpses of many returning birds "on sallows in the windy gleams of March" and we look confidently for more as April comes in "with hey, ho, the wind and the rain." It is without doubt our most capricious month.

The Artist and I were walking down a country road one day in early April and our heads were bent low to a gusty wind. Even the Crows were making heavy

75

weather of it aloft. Or as Tennyson put it so pictur-esquely: "The rooks were blown about the skies." High winds bother small land birds. They skulk in gullies or linger in thick cover or seek the lee of natural wind-breaks. Sea birds, on the other hand, often seem to enjoy a gale. I told the Artist how I found Ring-billed Gulls riding a half hurricane in Florida with ease and nonchalance and, on another occasion, I came upon a flock of Lesser Scaups actually playing in the surf dur-ing a northeast storm that included a driving rain. The ducks came tumbling onshore like a lot of schoolboys and time and again rushed back in for another ride on the roller coaster. It was evident that they enjoyed it and were doing it deliberately.

The sea is a distasteful subject to the Artist. He had made only one sea trip in his life and it made him very seasick. Possibly to stop any further mention of a pain-ful topic, he broke his stride to haul out his magnifying glass and then stopped to inspect a spray of Red Maple blossoms with his glass.

"I shouldn't do this," he said, shaking his head dole-fully. "It's too discouraging for a painter. Look at the colors in this blossom—the design—the delicate shading —wonderful! A man could spend a lifetime with paint and canvas and never come close to producing a single masterpiece like this one little blossom. Yet this one tree produces thousands of them. What is that thing you quote about a man being ridiculous?"

I took it that he meant a short sentence from *The Narrow Corner* by Somerset Maugham: "Life is short,

Nature is hostile, and Man is ridiculous." But with regard to the flower of the Red Maple over which he was enthusing, at least it attracted some public attention, whereas the less conspicuous but equally marvelous flowers of many other trees were ignored by most citizens. The elms, the ashes, the oaks—all the trees of our forests—had flowers of some kind—and fruit, too—but at this point the Artist was no longer listening, he was singing. With his finger pointing toward the stout trunk of a leafless Red Oak a dozen feet away, he had let go with:

"A song to the oak, the brave old oak,
Who hath ruled in the greenwood long;
Here's health and renown to his broad green crown,
And his fifty arms so strong."

He often breaks into snatches of opera while we are on our walks. No great harm is done, though sometimes he frightens off the birds or startles innocent bystanders who are not prepared for his lyrical outbursts. I interrupted his song by hinting that he was, in a measure, barking up the wrong tree this time. Henry Fothergill Chorley, music critic of the *London Times* a century ago, was chanting the praise of some English species of oak, not an American tree, when he wrote that familiar song.

"Same family," said the Artist stubbornly. Same family once removed would be a better way of putting it. But the interruption stopped the flow of song and the walk went along in silence—save for the sound of the wind in the trees—until we came on the glistening hulk

77

of a once lordly but, alas, long-dead Chestnut tree at the roadside. These relics still stand along our roads and in our woods, melancholy reminders of joyful days of old in this region when the great trees opened their bristly burrs with the velvety lining of a bright October morning and let their treasures drop to the ground.

"The first poem I learned in school," said the Artist, "was 'under the spreading chestnut tree.' I remember being called up in class to recite it. We had wonderful Chestnut trees near my home when I was a boy in Massachusetts. Well, you have to look hard to find a Chestnut sapling alive now—or a village blacksmith, for that matter."

It was evident that the Artist was out on a limb again poetically. The tree of the famous poem by Longfellow was not the bearer of the delicious fruit that was gathered eagerly by schoolboys in lost Octobers of long ago; it was not the true Chestnut, the noble *Castanea dentata* that had died a lingering death all over New England in the past half century. The Longfellow tree had a local habitation and a name. It stood in front of a blacksmith shop just a few steps off Harvard Square in Cambridge and it was officially identified by a member of the poet's family as a Horse Chestnut, or *Aesculus hippocastanum*. But since the Artist had brought up two poems about trees, I felt privileged to offer an excerpt from one of my favorite poems on the same subject. It was from Kipling and it ran like this:

Of all the trees that grow so fair
Old England to adorn,

Greater are none beneath the Sun
Than Oak, and Ash, and Thorn.
Sing Oak, and Ash, and Thorn, good sirs,
(All of a Midsummer morn!)
Surely we sing no little thing
In Oak, and Ash, and Thorn!

"Thorn?" said the Artist. "Did he mean the Black Locust?"

No, he meant one or all of the English haws or hawthorns, the members of the genus *Crataegus,* any species of which had its sisters and its cousins who are reckoned up by dozens in Gilbert and Sullivan fashion in this country as well as in England. Black Locusts, incidentally, are not native to England. William Cobbett took some from this country and set them out in England with the idea that the growth eventually could be turned into bean poles, fence posts and other useful articles at a profit, but I never heard that anybody made an immense fortune at it.

"I like the haws," said the Artist. "They have beautiful white flowers in the Spring—the leaves and fruit are bright red in the Autumn and add color to the pasture land—and the haw apples or hips or whatever you call 'em are fine food for the Ruffed Grouse in Winter. You can usually find the snow under the haw bushes covered with the feathered footprints—pretty as a picture—where the Grouse have been jumping up to pick at the haw apples still on the bush. They look good enough for anybody to eat."

There are other birds and some animals that are not

79

above nibbling at the haw apples in the dead of Winter. Reynard the Fox has been known——

"Excuse me for interrupting," said the Artist, "but I want to tell you about the last time I nibbled at a Ruffed Grouse. I was riding horseback along the wood road above our place in the Berkshires in the early morning of the hunting season and I looked up and saw a hunter in a tree. A big White Oak, it was, and the hunter was sitting on a limb over the road with a Ruffed Grouse in his claws. The hunter was a Barred Owl. I didn't know then what it was holding, but the owl flew off and dropped the object as I came up on horseback. I saw where the object had caught in a small tree, so I got down and salvaged the booty. It was a Ruffed Grouse, fresh-killed, and the owl had eaten the head off it. Otherwise the Grouse was all there and untouched, so I took it home, plucked it, cooked it and ate it."

It's a curious family habit of the owls to eat their prey *da capo*. At least that seems to be the custom of the owls in our region as far as we have observed. We found Pheasant and Rabbit remains and, on one occasion, a Scaup corpse indicating that the owl captors had started dining by taking, as the old song had it, "a little bit off the top." At this stage of our walk we had reached a point of vantage from which we could survey the best part of our favorite swamp, which we did, and spotted some Wood Ducks swimming quietly where there was a patch of open water guarded by dried reeds. The Artist is all agog when the Wood Duck returns in the

Spring. He says that the male Wood Duck, with its striking pattern of glowing colors, is the most beautiful native bird of North America.

There could be quite an argument about that. Probably it would come down to a matter of taste or personal opinion. There are wood warblers that have marvelous patterns and flashing colors. The orioles and tanagers would poll a flock of votes in a beauty contest. Why, there are dozens of woodpeckers of dazzling hues and picturesque patterns hammering away at trees from Maine to Oregon and from the Gulf of Mexico to the Arctic Ocean.

"Not one of 'em can sing worth a nickel!" said the Artist stubbornly as he edged out among the reeds—the wreckage of last year's growth of Broad-leaved and Narrow-leaved Cat-tail, mostly the former—to peer about in search of Virginia Rail, a task somewhat like that of looking for a needle in a haystack. Rails are the most modest and retiring of birds. They positively hate public notice and slink furtively away as any intruder splashes through their watery haunts. But I didn't let the Artist's snide remark about the singing of woodpeckers pass unnoticed. No genus or species of bird had everything in its favor. There is no direct relation between sight and sound in the feathered kingdom. The gaudy Peacock and the glittering Pheasant emit ear-piercing squawks. Woodpeckers had more than their voices to recommend them. They were sacred birds to the Romans because, according to Plutarch, they brought a variety of food to Romulus and Remus to

81

supplement the monotonous milk diet supplied by the famous wolf. The Virginia Rail that the Artist was vainly stalking—he was over his shoe tops in water by this time—had a hard rasping voice when it uttered its love call. The Wood Duck, so highly held by the Artist as a thing of beauty, had only a whispering whine or a purring note as a serenade to its inamorata.

"There may—for a change—be something in what you say," admitted the Artist as he splashed back toward solid ground again. "The Hermit Thrush is our finest singer and about as dull-colored a bird as we have in all New England. They tell me the Nightingale isn't much to look at, either. Maybe it works that way—a law of compensation—what a bird lacks in color it makes up in song. That would be fair enough."

A nice thought, but the rule doesn't run that way or any consistent way. The Baltimore Oriole is a beautiful bird and a fair singer, too. The Rose-breasted Grosbeak is strikingly handsome and its rolling warble is one of the most pleasant songs to be heard in the greenwood. Our eastern thrushes are quietly garbed, but all of them have lovely voices. Our sparrows are, for the most part, brownish birds with streaked or unstreaked breasts. They vary widely in musical ability. Some—the Vesper Sparrow and the Fox Sparrow, for instance—have melodious songs and others—like the Savannah Sparrow and the Grasshopper Sparrow—are little better in the operatic line than so many beetles.

Apropos of the Grasshopper Sparrow, I never had seen the bird until one day in Dutchess County when,

while thumbing Chapman's *Handbook*, I read that the
Grasshopper Sparrow often was overlooked where it
was common because its voice was so much like an insect that nobody looked in that direction to find a bird.
I was struck instantly with the remembrance of sus-

picious "insect songs," grasshopper or something else, in the meadow below the farm buildings, so I put down the book, picked up my field glasses, went out into the meadow, heard the suspicious thin buzzing sound, tracked it down and found it was coming from a Grasshopper Sparrow perched atop a green spray of Curled Dock. Within an hour I found a dozen more in the cutover hayfields and on fence posts. Since that time I can distinguish the song of the Grasshopper Sparrow at a fair distance when I am riding through farming country in an auto, or even in a railroad train.

But as a group, I think I would give the singing prize in our region to the thrushes. The Wood Thrush is the friendliest of them and its fluted song is familiar to suburbanites as well as countryfolk. It comes readily to suburban lawns and birdbaths and, though not as bold as the Robin—a close relative, by the way—it makes itself at home in our dooryard shrubbery and shade trees. The Veery (or Wilson's Thrush) long has been one of my favorite singers. It is primarily a bird of the swamps and wet woods and it was in a swamp in Dutchess County that I heard it as a boy and named the mysterious singer a "reed bird" because it seemed to be playing the melody on a reed instrument. Its rippling, circling, descending notes fascinated me and often I lingered long on the fringe of the swamp listening to the hidden singer in the greenery of early Summer. I still love the song, perhaps partially for "auld lang syne," remembering the quiet evenings when I sat on the ridge that overlooked the "Pine Forest" and listened to the Veery

84

singing near the brook below as a glorious twilight, heartbreaking in beauty, faded slowly into darkness on the western hills so many years ago.

The Olive-backed Thrush, which doesn't breed in our neighborhood but often favors us with music during migration, has a song somewhat like the Veery except that it spirals upward instead of downward and, on migration at least, is somewhat faint, hesitant and tentative compared with the confident and continued artistry of the Veery. But the songs of the Wood Thrush, the Olive-backed Thrush and the Veery are set to a definite pattern or motif that the birds follow over and over again, whereas the Hermit Thrush has a wilder and more varied song, with a margin for improvisation in each delivery. The Hermit Thrush is definitely a bird of the wildwood and its song is more delicate and ethereal than any of the others. It has a liquid quality suggesting

Music that gentlier on the spirit lies
Than tired eyelids upon tired eyes.

I feel that this might have been the thrush—though, of course, it was far from it—whose voice seemed to say to John Keats as set down in his famous sonnet:

Oh, fret not after knowledge, I have none,
And yet my song comes native with the warmth;
Oh, fret not after knowledge; I have none,
And yet the Evening listens.

Which reminds me that I received a letter from my friend, the Poet, on the same subject. Some years earlier the Poet had been the star fullback on a highly successful Vanderbilt football team, but at the time he wrote the letter he was the dignified head of the Department of English Literature (and boxing coach on the side) at one of the smaller ivy-covered colleges of New England. (Amherst, to be exact, which exposes my friend the Poet as David Morton, of whose friendship I am inclined to boast a bit.) His communication ran as follows:

The reports of your excursions to the woods with Herman the Magician, the Artist, the Astronomer, et al., contain too much about sight and not enough about sound. Poets concentrate on the songs of birds and leave the feathers for the scientists to pick over and sort out. Chaucer had his "smale fowles maken melodye." Shakespeare is full of bird songs with little mention of plumage. The famous *Ode to a Nightingale* by Keats is devoted entirely to the matchless song of the "immortal bird" that hungry generations cannot tread down. T. S. Eliot's nightingales "sang within the bloody wood when Agamemnon cried aloud" and so forth. Browning's wise thrush even sang each song twice over for poetic reasons as stated. I leave with you this poetic thought—and this bit of verse I concocted at the dawning:

> *I lie here, listening, now*
> *To the first bird calling,*
> *From blossoms on the bough*
> *Or blossoms falling,*
> *Through rainy dawn in Spring,*
> *I lie here, listening.*

P.S. Of course, even in pensive mood I couldn't lie there long. Not in New England! Early rising is a tribal custom in these parts. It suits me well enough.

Certainly it should suit him well enough. Every poet in the rural areas—every Singer of Persephone—should be regularly "brushing with hasty steps the dews away to meet the sun upon the upland lawn." But I am in favor of his attitude on bird songs. When World War II was at its bitterest hour I was involved in a transatlantic broadcast and cuckoos came into the conversation. We were talking back and forth across the Atlantic—three of us in New York and three in London—and one of the Londoners, a British officer just back from the fierce fighting in North Africa, said that the loveliest sound he heard when he reached home was (this was in late April or early May) the familiar call of the Cuckoo. He even went so far as to quote the two lines of the oldest song—for words and music—of English record:

> *Sumer is icumen in,*
> *Lhude sing cuccu!*

I take it that they really do feel that way about the common native Cuckoo (*Cuculus canorus*) in the British Isles, and perhaps it is best explained by Elizabeth Barrett Browning in one of her *Sonnets from the Portuguese:*

> *Remember, never to the hill or plain,*
> *Valley or wood, without her cuckoo-strain,*
> *Comes the fresh Spring in all her green completed.*

The Artist begins to fidget by the middle of April. He wants to be off to his stone house in the Berkshires where he paints outdoors for half the year. But before

87

he goes we always make three or four close inspections of the Van Cortlandt swamp for rails of any kind. It is almost useless for us to hope for success without the Artist. There is no new growth to hide the birds in the swamp in April, but the dried cat-tails and Phragmites of the previous season make good cover and rails are the champion skulkers of the bird world. Herman the Magician says that they are so clever they can hide behind a leaf shadow. But they can't hide from the eye of the Artist. If they are in the swamp and make a visible move while he is looking over the area, he will spot them every time.

Of course, rails aren't much to look at even when you see them, but you see them so seldom that it's a triumph to find one or more in a morning at the swamp. You can hear them easily enough in the courting season. Our commonest rail in the swamp is the Virginia Rail, whose voice has the unrefined quality of those ratchet-gadgets that revelers grind to make more noise at football games or New Year's Eve parties. But hearing the Virginia Rail is one thing and seeing it is another. I have heard the harsh, grating notes of the bird only a few yards from the end of my nose in the swamp and would have to confess

Even yet thou art to me
No bird, but an invisible thing,
A voice, a mystery.

But when the Artist is along, it's a different story. His keen eye catches them if they are anywhere in sight as they slink through the stalks of the cat-tails at water

88

level. Now and then we are lucky and have a clear view of a rail crossing from one wet sector to another by way of a dry patch or a raised path across the swamp. On one occasion the Artist and I stood just a few feet apart in the swamp and held perfectly still as a Virginia Rail walked between us, seemingly not noticing us at all. We have seen the King Rail and the Sora in this swamp at intervals, but the Virginia Rail is a regular and sometimes abundant Summer resident.

Some years ago the Florida Gallinule nested there and old birds with young following them about were common sights through the Summer, but for some reason—

perhaps they were harassed by boys with .22-caliber rifles—they stopped breeding there and we have seen them in recent years, as we see Coots in the same area, only as migrants. One day while walking on the fringe of the Henry Hudson Parkway where it passes through Riverdale, I picked up a dead Florida Gallinule. It was in the Spring migration period and the bird was in fine plumage that was hardly ruffled. There were no signs of teeth or claw marks on the corpse. Evidently it had killed itself by hitting a wire or some other object that it didn't see while migrating northward at night. I showed the dead bird to a couple of policemen who came riding along in a squad car. They were astonished that such an odd-looking bird should turn up on their beat, dead or alive. They both swore they never had seen the like of it.

One April day we left the swamp with the Astronomer in the lead and swung up over the ridge to the westward to reach the Hudson River. We had to go through the residential district of Riverdale on the way and by an act of trespass on a lawn we learned something that startled us in a botanical way. Even the Astronomer, who studied botany at Johns Hopkins, hadn't noticed it in his previous travels. We had been peering at the flowers of maples, elms and other native trees under our pocket magnifying glasses. We are always hauling out these magnifying glasses to have a larger look at the details of flowers of all kinds. It's a cheap way of having a vast amount of amusement. Some of the commonest little flowers of our roadsides and fields look, under magnification, like the most expensive beauties of a Park Avenue florist shop.

This day we saw a clump of cultivated yew of some kind on a broad lawn in front of an imposing private residence, and amid the dark green foliage of the yew we could see small flower clusters that were pale pink in color. This was another item to put under the magnifying glass. The owner of the lovely lawn wasn't in sight to shout "Avaunt, caitiff!" so I blandly trespassed on private property and bent down to examine the flowers of the yew under my little glass. As I shifted my scrutiny from one flower cluster to another, my sleeve brushed the yew branches and what looked like a cloud of smoke shot out from the bush. I pulled back in astonishment—and another puff of smoke sprang out. The Artist and the Astronomer "looked at each other with a wild surmise." The Astronomer regained his poise first.

"Pollen!" he said quickly. "Regular clouds of it. I've read all about it, but I never saw it happen before."

It was plain that, in bending over the bush, I had brushed some flower-laden branch in such a way that it sprang back when released and threw off the pollen grains in a little white cloud. Just to see how far this could go, I reached down and shook a branch of the yew. The green bush sent up clouds of pollen in all directions. It was like a miniature volcano in action. And then I suddenly recalled two lines of Tennyson, the full meaning of which I never before had grasped. They are from the *Idylls of the King*—the words of Brother

Ambrosius to good Sir Bors when the knight, returning from his quest for the Grail, rested for an evening at a monastery:

> *O brother, I have watched this yew tree smoke*
> *Spring after Spring for half a hundred years.*

So our Sunday saunter and act of trespass gave us a better understanding and deeper appreciation of Tennyson as a poet and a botanist. As a matter of fact, Tennyson knew a great deal more of botany and birdlife than did William Wordsworth, the poet famed for his verse about Nature and what stems from it in human thought.

On this same trip the Artist saw a Slippery Elm and insisted on cutting himself a small segment of the inner bark for chewing purposes. He explained that he was once a barefoot boy. No explanation was necessary. All of us had chewed Slippery Elm in our younger days. As for the Artist, he goes about collecting bits of Sassafras root, Winter and Summer, and sticking them in his pocket. He says he likes to chew the root while he is painting. He also takes an occasional glance out the window while painting, and if he sees an uncommon bird in the sky, the trees or the shrubbery, he puts down his brush and palette, grabs up his field glasses and runs out of the house to get a better view of the distinguished visitor.

The last birds he checks on before leaving for the Berkshires "when April melts in Maytime" are the swallows that swarm over the reeds and waters of our favorite swamp at that season. The common—even abundant—swallows there in April are the Barn Swallow, the Tree Swallow and the Rough-winged Swallow, but we catch sight of a scattering of Bank Swallows and Cliff Swallows and one fine day we saw all five species together. The Artist puts out boxes for the Tree Swallows that nest with him in the Berkshires year after year, and he feels especially friendly toward that species, but he loves all the swallows for their grace and beauty in flight and it grieves him mightily that the Purple Martin has disappeared almost completely from most of the territory over which we roam and throughout which this friendly bird was once a common Summer resident.

7

Now, my fairest friend,
I would I had some flowers o' the Spring that might
Become your time of day; and yours, and yours.
WINTER'S TALE, ACT IV, SC. 4

May is the feverish month with us because it marks the high tide of the warbler migration in our territory. Myrtle Warblers stay with us through the Winter where they find Bayberry bushes or some other regular food supply and occasionally we find a stray warbler of some other species on our rambles during the cold months. The migrant Myrtle Warblers and Pine Warblers and Yellow Palm Warblers and one or two more of the hardier species come along any time between the last week of March and the end of April, and the stragglers of the migrants that pass through our region may drift into our view as late as the first week in June, but the great tide of the warbler migration sweeps over us in the second and third week of May, the fortnight that

94

brings something approaching a prolonged state of delirium to bird enthusiasts.

The late Dr. Frank M. Chapman, the ultimate authority on the subject and author of *The Warblers of North America,* wrote that the warblers are "our most beautiful, most abundant and least known birds." They are, indeed, surpassingly beautiful as a group and certainly they are abundant in our woods, fields, orchards and hedges during the Spring and Autumn migration periods, but they are small birds—about four to seven inches in length all told—and they are rarely noticed except by those who look for them expectantly and view them each time with renewed delight.

The trickle of warblers in April is followed by the colorful wave in May. There are, according to Chapman, about fifty-five species in North America. Some three dozen species are regular or occasional visitors in our territory either as breeders or migrants, but it takes a field expert on the move all the time—with a dash of luck added—to "log" all the possibilities. We have our good days and our poor days with our "warbler count," our good years and our bad ones. Occasionally we spot a rare species amid great jubilation, but more often it is the other way around: we miss seeing a single individual of some common or fairly regular species. Possibly we would have more species on our seasonal lists if we had the Artist with us more often in May. His keen eye might catch a few that we others miss, but I'm not sure of that because I, for one, hunt the warblers largely by sound. I hear their songs in most cases before I start to look for the singers.

Here's a list of twenty-five species that we see with fair regularity either as Summer residents or migrants in our region: Black and White Warbler, Worm-eating Warbler, Blue-winged Warbler, Nashville Warbler, Northern Parula Warbler, Eastern Yellow Warbler, Magnolia Warbler, Black-throated Blue Warbler, Myrtle Warbler, Black-throated Green Warbler, Blackburnian Warbler, Chestnut-sided Warbler, Black-poll Warbler, Northern Pine Warbler, Northern Prairie Warbler, Yellow Palm Warbler, Oven-bird, Northern Water-thrush, Louisiana Water-thrush, Northern Yellow-throat, Yellow-breasted Chat, Hooded Warbler, Wilson's Warbler, Canada Warbler and Redstart.

We know most of their songs—though some of the common species with what I call the "switch-switch-switchy" type of song confuse us at odd times—and have the birds identified by ear before we can locate them with our field glasses. But some seasons we can't find any Blackburnians in the migrating horde—or it may be the Hooded Warbler that escapes our observation. Other seasons we find Blackburnians almost in abundance and find Hooded Warblers plentiful in the deeper tangles of wet woods in May. The Hooded Warbler, one of my favorites, has a ringing song that, once it is known, can be recognized at a furlong distance on a quiet day in the woods.

It was in the lush undergrowth of a well-watered patch of woods in Westchester in mid-May that Herman and I began a botanical dispute that lasted some weeks. I was checking over the warblers that were plentiful on

96

such a morning in wet woods and the Astronomer and Herman were searching the ground for flowers, including the Yellow Lady's-slippers that were still in the bud at that time. Every now and then as we worked slowly through the underbrush, Herman would point to some feathery foliage and say "Black Cohosh" or "White Baneberry." I looked where he was pointing each time and couldn't notice any difference between what he was designating "Black Cohosh" and what he was calling "White Baneberry." I said as much and one word led to a thousand more, some of them being the Latin names of plants that were exchanged violently at four paces.

Herman said that what he meant by "Black Cohosh," also called "Black Snakeroot," was *Cimicifuga racemosa* to the botanist, and what he called "White Baneberry" was *Actaea alba* to the botanist. He said the woods around us were full of both and he waved his arms in lordly fashion in all directions by way of proof. The Astronomer, who was enjoying the debate, didn't step in to settle the matter because that would have ended the fun. He let the argument run riot.

I challenged Herman to show me how he distinguished the foliage of his *Cimicifuga racemosa* from that of his *Actaea alba,* and his only answer was that the *Cimicifuga racemosa* was much the taller of the two. That didn't seem like a scientific distinction to me and I insisted that what he had been pointing out as two different species of plant were, for my money (I think I mentioned as much as two dollars, current coin of the realm), one and the same species at different stages of growth. It's too bad that the Astronomer didn't fetch me a stunning blow on the sconce at that precise moment, because I was eminently correct up to that point. But, like many an orator, I kept on talking and spoiled everything. I said I would name the species. I had my choice of *Cimicifuga racemosa* or *Actaea alba*. I made a bad guess and picked *Actaea alba*. Once I had picked it, I had to stick to it defiantly and declare that there wasn't a *Cimicifuga racemosa* in the lot—the wood lot.

With the Astronomer resolutely holding his peace so that he could laugh without prejudice as the debate progressed, there was nothing to do but wait for the plants to grow up and prove one of us right and the other wrong. The flowers and fruit of the plants in question have wide differences in appearance, though the plants are close relatives botanically. The *Actaea alba* is the species that produces the fruit called "dolls' eyes" because of the remarkable resemblance of the black-tipped white berries to the artificial eyes that are put in the largest and most expensive dolls manufactured for the juvenile carriage trade, so to speak. With their glowing purple pedicels, their smooth white skins and their black tips that are the "pupils" of the "dolls' eyes," a full stalk of these oval-shaped berries is something to gather in the early September woods and bring home as a natural *objet d'art*. I once found a stalk with thirty-six perfect berries on it and carried it 160 miles to display it before admiring friends.

But the point at issue in the mid-May woods was whether or not I recognized the plant in its salad days.

98

The following week we were back in the same woods again and we looked to see how our debate was flowering. I did not like the way things were coming along, but it was still too early to make a definite decision. Two weeks after the firing started we were back again on the scene and I was sorry that the matter ever had been mentioned. By this time it was apparent from the way the pale green flowering stalks were shooting upward that those woods were going to be full of *Cimicifuga racemosa,* and my contention that they were *Actaea alba* was fast becoming a blooming farce.

I gave up officially before the flower spikes burst into their feathery white blooms but I didn't surrender unconditionally. I retreated to the point where I had insisted that all the early foliage Herman offered for inspection was of one and the same species, not two different species, as he had asserted. I merely went wrong when I made my guess as to which flowering plant it was. But Herman the Magician had maintained boldly that he had been pointing out *Actaea alba* as well as *Cimicifuga racemosa* in the incipient stage in the woods. All his alleged *Actaea alba* either pulled up roots and walked away or grew up to be *Cimicifuga racemosa,* because we didn't find a single stalk in those woods that grew to maturity as *Actaea alba.* That was Herman's minor error. I committed the more glaring error, and Herman made the most of it for many months in perambulating conversation when we were in the field.

More matter for a May morning was the corpse of a

Gray Fox that we found in a lonely glen a few miles from the starting point of our botanical dispute. Sunlight was streaming down through the young foliage of the trees around us and warblers of many species were fluttering about, high and low, singing and feeding as they worked their way northward on migration. The weather was marvelous and the scene was beautiful. We were feeling on that flowery morning—the floor of the woods was sprinkled with Narrow-leaved Spring Beauties, Wind-flowers, Dog's-tooth Violets and other seasonal bouquets—as Wordsworth must have felt when he wrote:

> *Then sing, ye Birds, sing, sing a joyous song!*
> *And let the young Lambs bound*
> *As to the tabor's sound!*
> *We in thought will join your throng,*
> *Ye that pipe and ye that play,*
> *Ye that through your hearts today*
> *Feel the gladness of the May!*

We were looking in particular for Yellow Lady's-slippers and were working our way up the rocky bed of a somewhat skimpy stream that loitered down through the woods. We had found patches of Showy Orchis on the moist slopes and rejoiced to see them, but the Astronomer needed some good color pictures of Yellow Lady's-slippers to fill out his slide collection and we were on the hunt for a handsome patch along the watercourse. Yellow Lady's-slippers like to keep their feet wet.

Herman the Magician is our orchid spotter and he

located a perfect patch for the Astronomer's camera: fourteen Yellow Lady's-slipper plants in bloom together, some of them carrying two flowers each. The Astronomer and Herman had to engage in a bit of amateur engineering to fix things so that a close-up picture could be made without forcing the photographer to kneel down in mud and water. They gathered fallen timber and made a trestle on which the Astronomer could lie as he operated his camera. I was listening to the familiar songs of warblers around me and watching the amateur engineers at work when, amid the warbler concert, I heard the buzzing song of the Worm-eating Warbler. It is far from a melodious song and the warbler is one of the plainest of that entrancing family, but every Spring I like to clap my eyes on at least one individual of each species, migrant or Summer resident, and I was determined to track down this singer. It would be my first Worm-eating Warbler of the year.

I left Herman the Magician and the Astronomer kneeling on their trestle by the Yellow Lady's-slipper patch and headed up the steep hillside in the woods. The traveling was difficult because the way was blocked by fallen timber and huge jagged rocks, but the warbler kept buzzing and I kept scrambling upward in the direction from which the sound was reaching me. When I judged that I was close enough to make a scrutiny, I braced myself against the trunk of a tree on the steep slope and swept the branches and foliage overhead with my glasses. The Worm-eating Warbler, more often than not, is hard to find because it has a habit of sitting

102

in one place for an extended period while singing. Most warblers sing "on the run," so to speak. They hop and flit about through the foliage, feeding and singing by turns, in sight one moment and lost to view the next. But their movements catch the eye. Not so with the Worm-eating Warbler on many occasions—and this was one of them. I developed a real pain in the neck looking upward for twenty minutes, but by that time I was determined not to be baffled by the bird and I kept peering overhead until I found it perched on a branchlet on the outer fringe of a Sugar Maple about forty feet directly over my head.

I was still watching the perched bird and listening to its song when I heard a shout from Herman below in the glen. The pictures had been made and, going up the watercourse, Herman had come upon the skin and bones of a late and probably unlamented Gray Fox. Herman conducted an inquest on the spot and, after examining the body, decided that the victim had died of gunshot wounds at the hands of some person or persons unknown, possibly a farmer who heard a flutter of excitement in his poultry yard, grabbed a shotgun and fired with accuracy when he caught Reynard "in flagrante delicto." The wounded intruder, as Herman reconstructed the tragedy, escaped from the scene of the crime but reached the woods only to die of wounds in this lonely spot.

We are not, as a group, precisely in favor of purloining poultry, but somehow we regretted that the Gray Fox had been cut down. An occasional glimpse of a

Red or Gray Fox in the woods or fields is an adventure and we often speculate over the fox tracks that we find in the snow—where this chap was going and upon what business he was bent by Winter moonlight along the shore line of a frozen lake. We mourned silently for a minute or so by the remains of Reynard and then strolled ahead through the woods. The "gladness of the May" had received a temporary check but it was not of long duration because, as we went over a ridge, a Ruffed Grouse whirred off through the trees and a few yards further along we found a gorgeous clump of Pink Azalea in bloom. We bent over to inhale its perfume, which is one of the rites of Spring with me. The sight of the Ruffed Grouse and the odor of the Pink Azalea lifted our hearts again and we went our way rejoicing.

A few days later a former left fielder for Yale sent word that he had a family of Red Foxes in the field next to his orchard in Westchester County. We went over to investigate and found the fox earth at the base of a White Pine in a fallow field sprinkled with scrub growth of various kinds. We found fur and feathers scattered around the entrance to the tunnel and concluded that the vixen and kittens had been dining on Crow, Blue Jay, Woodchuck, Rabbit and Pheasant. A little further off we found signs that a young Skunk had been added to the menu, perhaps as a morsel of delicate flavor for dessert.

The former left fielder of the Yale varsity nine had an Airedale named Old Nick that followed us into the field. Old Nick poked his nose down the entrance of the fox earth and the stump of his tail wagged furiously as he whined and pawed and tried to crowd himself into the hole. Nothing came of that, of course. Like Hamlet, Old Nick was "fat and scant of breath." We left him hard at work and went into a nearby patch of woods to look for Spring flowers and birds. When the dog came galloping along to join us a few minutes later he was carrying half a Musk-rat in his mouth. Evidently he had found a meat cache of the fox household and had rifled it ruthlessly. Old Nick seemed quite proud of his booty. Perhaps a dog should be allowed to swagger a bit when he has filched something from a fox.

The Purple Trillium was profuse in this patch of woods and whole patches of the ground under the trees were covered with Dog's-tooth Violets. There were banks that were white with the Narrow-leaved Spring Beauties and in moist pockets on rock shelves there were exquisite white chalices of Bloodroot. Here and there in the open spaces we found the High-bush Blueberry hanging out myriad clusters of bell-shaped flowers. From the top of a tree in the middle distance a Brown Thrasher was giving a concert. Out of the branches of a Hemlock close at hand came the bubbling song of the Ruby-crowned Kinglet, a dainty offering from a dainty bird.

We heard the "chip-churrr!" of a Scarlet Tanager male and soon found the bird, a spot of flame amid the light green foliage of a tall tree. Not far away a Rose-breasted Grosbeak was warbling steadily, and from the underbrush just ahead of us there came the "Teacher-

teacher-TEACHER!" of the Oven-bird. What a powerful voice for such a small singer! And what an odd bird it is, a member of the warbler family that walks sedately on the ground instead of flitting lightly through trees and shrubbery, and builds its nest on a slant in a hollow on the ground, by which habit it derives its name of Oven-bird from the resemblance of its nesting site to a miniature oven. For those who read Thoreau and remember his notes about the mysterious "night warbler" whose voice rang out under cover of darkness in the woods and whose identity he never could learn, ornithologists believe the culprit was the Oven-bird, which has a rather wild chatter that it turns loose in the dusk or even late at night in the woods. The Artist and I have heard it on more than one occasion in the woods that cover the high hills of western Massachusetts.

On May 10 our group of strollers took to a high spot in Westchester County, a ridge covered with great Hemlocks. I call them the "Gustave Doré Hemlocks" because they look like the Doré illustrations in an edition of Dante's *Inferno* over which I pored as a boy. We approach this ridge through a small swamp at the inlet of a large pond or small lake. When we go through there in April we hear the resonant calling of the Pied-billed Grebe. On this May morning we did not hear the Pied-billed Grebe, but there were migrating warblers singing all about us as we crossed the swamp and began the steep climb to the ridge adorned by the noble Hemlocks.

Just at the point on the slope where the Hemlocks

take over from the hardwoods, Herman the Magician bent over and pointed to a clump of Showy Orchis coming into bloom. On higher ground, on the floor of the forest where the great Hemlocks towered over us, we found the Pink Lady's-slippers in profusion and almost all of them in what I enthusiastically pronounced the pink of perfection. There also we found the delicate fronds of the Maiden-hair Fern as well as many other species of fern such as the Marginal Shield Fern, the Christmas Fern, the Cinnamon Fern, the Interrupted Fern, the Sensitive Fern, the American Shield Fern, the Common Polypody, the Ebony Spleenwort—not all in one spot, of course, but scattered over the terrain from the swamp to the top of the wooded ridge. The region is a wonderful hunting ground for ferns alone. Along one shoulder of the ridge, in a little hollow or shelf in the steep slope of the wooded hillside, we found —as usual—glistening green patches of the Shining Club-moss. All along the trail we had seen the creeping greenery of the Partridge-berry and the Ground-pine.

We were approaching the area where we had found the nest of the Great Horned Owl when Crows began calling in a loud chorus to the northward. Herman turned on his heel and left us immediately. We were catching one species of warbler after another in our glasses and having a great time, but Herman never bothers his head about such small birds. He is always after bigger game, and this time his hurried departure was in the hope of finding the Crows cawing excitedly around a Horned Owl. Crows are the best watchmen

107

of the woods and fields, with Blue Jays taking second place. Of course, practically all birds will raise a fuss if thrown into fright or peril by an apparent enemy of any kind, but Crows and Blue Jays seem to take it upon themselves to patrol their regions and deliberately harry other birds or animals that they consider dangerous intruders or impudent trespassers.

Crows dislike hawks and heckle them at every opportunity. They take after foxes, too. A fox usually has business reasons for wishing to slip unnoticed through the woods, along an old stone wall or across an open field, and he is much annoyed when he is spotted on such travels by these raucous-voiced, black-plumaged watchmen of the woods and fields. Crows raise a great clamor when they come upon an owl in a tree. They gather around the blinking owl and shriek malediction upon him. They take turns flying as close to him as they dare, which is not too close if it's a Great Horned Owl, because that bird has fierce talons and, unlike humiliated humans, will "eat crow" with gusto.

"We'll soon know what's up," said the Astronomer as the cawing continued to the northward. "If it isn't an owl—and a big one—Herman will be back in a few minutes."

We decided to work along the sidehill in that direction, looking over the migrating warblers as we went. The undergrowth through which occasionally we had to push our way consisted mostly of Yew, Moosewood, Witch Hazel and Mountain Laurel. Considering that Hemlocks *en masse* are not the best lure for all kinds of warblers, we did well on the ridge this trip. We saw, among other species, a Prairie Warbler and a Hooded Warbler. We usually find the Prairie Warbler in more open territory and the Hooded Warbler at a lower level in wet woods. But on migration a bird may be found almost anywhere in a region through which it passes. We saw two or three Blackburnian Warblers among the Hemlocks, and the sight of even one of these brilliant birds always enlivens a day afield. There were plenty of Northern Parula Warblers with the burnt-orange coloring laid on the breasts of the males in marvelous fashion. There were Myrtle Warblers all over the place and Black and White Warblers running up the trunks and around the branches of the Hemlocks in all directions. Redstarts were like flashes of flame as they flew about the Hemlocks, and the abundant Black-throated Green Warblers, too, stood out in glowing yellow against the dark background of the evergreen foliage. We saw some Nashville Warblers, birds that belong on the sober side of the family for color and pattern, but we also saw some Magnolia Warblers, which certainly belong in the top group for color contrast and striking costume design. We must have seen or heard a dozen Canada Warblers and as many Black-throated Blue Warblers. There isn't much color to a Black-throated Blue male, but certainly he is a neat dresser.

By the time we had taken note of this array of warblers—and perhaps a few more species that escape memory—we had come close to the conference headquarters of the cawing Crows and suddenly we saw a Red-tailed Hawk flap out of a tree and go off with a whistling cry. The Crows heard the hawk's note of alarm as it left the premises. They ceased their chatter and scattered silently.

"So it was a hawk they had after all," observed the Astronomer.

"Hawk nothing!" said Herman indignantly as he stepped from behind a tree just ahead of us. "That hawk just happened along. The Crows had a Great Horned Owl pinned in that big Hemlock until you fellows came crashing along like a flock of horses. Kindly make less noise hereafter."

I left Herman and the Astronomer to their own devices a week later and went to visit the Artist at his studio on a high hill in western Massachusetts. This is a part of the Berkshires, a region of wooded hills and little valleys. There is comparatively little open country. Because of the profusion of trees, the region is well watered. There is a trickling brook in every glen, a chattering stream dashing down every ravine and a little river winding through every valley.

The Artist already had his garden well under way when I arrived there on May 16. He always has a good garden and is particularly adept at raising luscious sweet corn. He had his bird boxes up and the usual tenants were installed, Bluebirds and Tree Swallows. He told me that they had been waiting for him to show up and flew about him, uttering querulous notes, as he nailed the boxes on their lofty supports again. By sundown of that day they had moved in and started housekeeping. While he was sorting out some old poles for the bean patch in his garden, a Pileated Woodpecker went past him in bounding flight. This is a bird that will make anyone sit up and take notice. It's about eighteen inches fore and aft—or up and down if it is hacking away on the trunk or branch of a tree in search of grubs concealed therein. It has a flaming red crest, below which it wears a loud costume of black and white. It really makes the chips fly when it goes to work with its whacking big bill, and it has a cry that may be heard half a mile with a favoring wind.

It was early afternoon when I arrived and, as we stood looking out over the valley, we could hear the river below us rushing heavily over the big boulders in its bed. There had been plenty of rain and the water was high. We knew that the trout brook to the north must be in good condition for fishing.

"Let's see if we can catch a mess of trout for supper," said the Artist.

It was a temptation, and Oscar Wilde said that the best way to get rid of a temptation is to yield to it. We hauled out old trout rods and pieced them together. Off we went, first up the hill and then along the wood road toward the brook. Now, there are various ways of fishing. It may be recalled that a famous Dickens character, one Jerry Cruncher (senior), went fishing at midnight in a cemetery and began to fish with a spade. He was a "resurrectionist" by trade, a body snatcher. Mr. Cruncher's style of fishing was illegal and indecent. No doubt of that. But barely a thin cut above Mr. Cruncher, as all proper followers of Izaak Walton know, is the scoundrel who uses worms for bait when he goes fishing for Brook Trout. A man who will do that is, by any civilized standard, lower than the worms he carries in his bait box. This is the law and the accepted judgment in correct fishing circles. Nevertheless, the Artist and I used worms for bait on this trip to the trout stream. We always do. We are outcasts, even Ishmaelites, but usually we get the trout. This time we didn't. We fished that lovely stream all afternoon until the sun was low in the west. There were wonderful pools that should have been teeming with hungry Brook Trout, but we caught only two—one apiece—of legal size.

"Vandals have been here," said the Artist slowly and sadly.

A "vandal" in such cases is a man who gets there before you and catches the fish you planned to catch yourself. But it wasn't a wasted afternoon by any means.

The warbler migration had just reached its peak in the Berkshires and there were a dozen species flitting through the underbrush around us or the trees overhead. A Louisiana Water-thrush, with many bursts of song, kept moving ahead of us along the stream. Several Blue-headed Vireos came peering at us to see how the fishing was going. A Northern Yellow-throat scolded us sharply for intruding on what looked like a capital nesting site in a high clump of weeds where the stream took a little turn in the open before plunging back into the woods again. Then, as we tramped home in the hush of twilight, we heard—one after another from the dim recesses of the trees all around us—the Hermit Thrushes sprinkling the leafy dusk with magic melody. It was truly wonderful, beautiful beyond speech, which is why the Artist and I said nothing as we walked homeward in the gathering darkness. I wish I could put in words all that I feel and think when I hear the Hermit Thrush in the woods at twilight, but the best I can do is to borrow from Victor Hugo a sentence out of *Les Misérables:* "Il y a des moments où, quelle que soit l'attitude du corps, l'âme est à genou."

The next morning we went for a tramp in the woods and the forest floor was so thickly carpeted with Spring flowers that, on my return to the Artist's studio, I sat down and drew up a list I long had planned but never had put on paper, a Cast of Characters for a Great Native Drama. Let somebody else write the Great American Novel. I had something more dramatic in mind and, if I could arrange it, I would have the curtain

go up to a flourish of Trumpet Vines. There will be
no charge for admission and I have yet to work out the
flowery details of any possible plot, but the May morning in the woods caused me to concentrate on the
problem of presenting, from among my acquaintances
met in the fields and woods over which we regularly
roam, a cast of characters to be described and billed
as follows in true theatrical fashion:

Dramatis Personae

Slender Gerardia, *a lovely heiress*   Agalinis tenuifolia
Smooth False Foxglove, *a designing villain*
                              Dasystoma virginica
Swamp Lousewort, *the villain's confederate*
                              Pedicularis lanceolata
Spider-Lily, *a night-club hostess*   Tradescantia virginiana
Virginia Stonecrop, *the old-maid housekeeper*
                              Penthorum sedoides
Daisy Fleabane, *the housekeeper's assistant*
                              Erigeron philadelphicus
Rose Pogonia, *a delicate child*   Pogonia ophioglossoides
Jimson Weed, *the village banker, a curmudgeon*
                              Datura stramonium
Calypso, *a dancing girl*   Cytherea bulbosa
Black-eyed Susan, *a farmer's daughter*   Rudbeckia hirta
Moth Mullein, *former outfielder for Detroit*
                              Verbascum blattaria
Prostrate Tick Trefoil, *a heavyweight boxer*
                              Meibomia Michauxii
Ragged Robin, *a good-natured tramp*   Lychnis flos-cuculli

114

Red Lobelia, *high school football captain*   Lobelia cardinalis
Herb Robert, *high school basketball star*
Robertiella robertiana
Dwarf Cornel, *a jockey*   Cornus canadensis
Trailing Arbutus, *a detective*   Epigaea repens

Costumes by Demeter & Persephone, Inc. Incidental music by Aeolus and his famous Wood Winds. Lighting effects by Jupiter & Son. Spring showers in Act I by J. Peter Pluvius. There will be a return engagement next year.

8

FALSTAFF
But are you sure of your husband now?
MISTRESS FORD
He's gone a-birding, sweet Sir John.
MERRY WIVES OF WINDSOR, ACT IV, SC. 2

It is time to record a small triumph of Herman the Magician, whom we call "the big bird man" because of his scorn of the smaller species and his enthusiasm for game birds, hawks, owls and eagles, most of which are on the large side. But on one occasion Herman outscored us in the smaller division, wherefor he rejoiced exceedingly.

Most of our rambles in May had led us into the woods and we had missed seeing any Yellow-breasted Chat or White-eyed Vireo. We like to keep up such acquaintances and we know a place where we can always find them in June. They haven't failed us for many years. We went there this day in June to find the Chat and the White-eyed Vireo. It's a wet meadow, almost swampy in the low spots, and heavily overgrown with Poison Ivy, Staghorn Sumac, Smooth Upland Sumac, Arrow-wood, Panicled Dogwood, High-bush Blueberry, Black Alder and Low Running Blackberry. Catbrier lurks wickedly in some of the thicker tangles and hangs like a barbed-wire barrier among the twisted limbs of the scrub apple trees that grow along the old stone walls of the fields.

This moist ground is the haunt of the Woodcock and we go there at dusk in April to see and hear the male bird's romantic performance culminating in the zigzag downward flight while a tinkling melody is sprinkled over its inamorata below. Here also in October we find the lovely Fringed Gentian, and in November the shining red berries of the leafless Black Alders are a sight to behold. But we were after other game this June morning and we had just entered the field when we heard a Chat somewhere amid the greenery that was thick along the stone wall. The Chat was in good voice. It gave off chuckles, caws, whistles and a noise like a dog barking at a distance. This bird is a ventriloquist as well as a mimic and a tease. It seems to enjoy being deceptive. You may hear it plainly but, though you look high and low, the singer manages to keep out of sight.

We were determined to have a look at the trickster and made plans to surround it and force it into view. The Medical Student circled to the far side of the wall. I moved closer to the source of the sound through

117

a patch of blackberry vines that tore at my ankles. The Astronomer was out on the flank to prevent a clean getaway in that direction. As the Medical Student started off, he disturbed a small bird that turned out to be a Blue-winged Warbler with food in its bill. Evidently its nest was somewhere in the vicinity, but we couldn't be diverted from our hunt for the Chat. After a scrutiny of ten minutes on his side of the wall, the Medical Student shouted that he had sighted the Chat in the thick of the tangle between us. While he was trying to direct my eyes to the spot by calling directions from his masked position on the other side of the wall, he flushed the bird toward me and I had a good look at it before it wheeled back into the thicket again, still chuckling, cawing and whistling to lead us on.

We were still stalking it for further inspection when we heard the decisive song of the White-eyed Vireo at the far end of the same stone wall. The Medical Student and I plunged in that direction on either side of the fence, pushing through the brush and the low-hanging branches of trees along the wall. For twenty breathless minutes the White-eyed Vireo led us a merry chase, much of it through luxuriant patches of Poison Ivy. Then it stopped singing and we never did catch a glimpse of it.

We wandered back to where Herman the Magician had been standing all this time, and he had a thing or two to tell us. In the first place, the Chat had come out on a dead limb not twenty feet away from him and there perched nonchalantly to give him a full-scale

Chat concert with flourishes. In the second place, the Blue-winged Warbler with food in its mouth had betrayed its nesting site at the foot of a clump of Meadowsweet. Herman proudly showed us the nest with five tiny young that could not have been long out of the shell, and then he stood back and asked expansively:

"Who says I don't know anything about small birds?"

June is, to be sure, the nesting month, and in our travels we often find young birds at home at this season. For the most part our discoveries are casual. Seldom do we start out to find a nest of any particular species. It requires detective work of a high order to find nests before that stage of family life when the young birds announce their presence by clamoring hungrily as the parents arrive on or near the premises with a fresh supply of food. The Autumn winds, stripping trees and shrubs of their leaves, expose dozens of nests that were unsuspected in the breeding season. But if the young are advertising their presence with outcries for food, a curious person merely has to track down the sound to find the quarry.

One way of finding nests before the young are audible is to make a full search of a region, inspecting every part of it from the ground up to the top of the tallest trees. This is an exhaustive procedure that we never bothered to put into operation in our territory. We merely strolled along and kept our eyes open. The nests of Baltimore Orioles are rather easy to find despite the way the parents try to conceal them in the thick foliage drooping from the outer fringe of American Elms, Tulip trees or whatever they may choose. These birds are conspicuous because of their colorful plumage and the male keeps singing cheerfully, so by sight and by sound it is easy to follow Baltimore Orioles in the nesting season until they lead the way to their nests. The male Rose-breasted Grosbeak sings in the vicinity of its nest and so does the male Scarlet Tanager. The females of these species are rather trustful of humans and often build nests in shade trees on suburban lawns. As for Black-capped Chickadees, Downy Woodpeckers, Bluebirds, White-breasted Nuthatches and other such birds, all that is necessary to find their nests is to scan the holes in fence rails, fence posts, telegraph poles and trees, dead or alive. Where a bird is seen going in or coming out of a hole, that place may be put under suspicion of containing a nest. The Flicker, I think, is the most amusing of the hole-dwellers of our region. It has a most entertaining way of popping its head out of the hole at the slightest provocation. In

120

fact, a standard method of learning whether or not a likely hole has a Flicker tenant is to rap on the wood below, whereupon the bird, if at home, will stick its head out of the hole and look around as if inquiring:

"Who's that knocking on my door?"

However, we don't like to disturb nesting birds. One reason is that it annoys the birds. Another reason is that, among the ground-nesting or low-building species, vermin may track humans to a nesting site with fatal results for the birds. Foxes, Skunks, hunting house cats and other such predators not only flee but sometimes follow humans in the wildwood, sniffing curiously or watching keenly to see what the humans are up to, and if a nesting site is visited regularly, one of these predators will follow the trail and find it.

The Artist tied a supply of unraveled hemp to a small tree in his dooryard in the Berkshires, and Purple Finches, Baltimore Orioles, Chipping Sparrows, and Goldfinches drew on it plentifully for nesting material. Just below my little cabin we found the nest of a Hermit Thrush at the foot of a Hemlock sapling. Fifty feet away we found the nest of a Brown Thrasher in the thick of a haw of some kind, and along the road a Kingbird pair had a nest atop an apple tree. Kingbirds like a nest from which they have a clear view of the sky. Aside from the Bluebirds and Tree Swallows that nest regularly year after year in the boxes he puts up for them, the Artist has Goldfinches nesting each year in a Sugar Maple by his door, Purple Finches at home in a Juniper on his lawn, and Phoebes that build under the roof of his side porch. The young Phoebes are always a sad mass of bird lice in their growing days, and sometimes the young succumb to the overwhelming numbers of the pests. There was a Ruffed Grouse that built a nest at the foot of a European White Birch along a stone wall just where there was a gap through which we went to a spring that had been stoned nicely by the Medical Student. The Artist found the nest and thereafter we avoided going through the gap in the wall to the spring, lest some animal should follow our tracks and disturb the nesting bird. All of us are great admirers of the Ruffed Grouse and we were hoping that the hatching would go ahead safely, but at the very last some animal killed the mother bird and dined on it. The Artist opened one of the eggs and found a perfectly formed chick, but the murder had been committed at night, and by the next morning the eggs were cold and the chicks within had died.

We found two other Berkshire bird families that, so far as we ever knew, had a safe and successful season. There was a boy named Stanley who did chores on a neighboring farm. Stanley was interested in birds and animals and often brought us news of what he came upon while at work or play. He told us we would find the nest of an Oven-bird if we would walk along the upper road to the pasture lot, cross the second lot to the west where three White Ash trees had been cut and were lying at the edge of the woods, go behind the fallen trees along a cowpath and fifteen feet inside the woods look under a Juniper on the left side of the path.

The Artist and I followed directions and found no nest. In fact, we didn't even find the Juniper. We went back for more detailed directions from Stanley and then retraced our steps. This time, as we went along cautiously, I saw an Oven-bird scoot out like a mouse from under a low Hemlock sapling. That was how we had missed it on our first search. Stanley had told us to look under a Juniper. We went down on our hands and knees by the Hemlock sapling and found the odd nest with the "side door," really built into the sidehill like an open oven, with four young birds therein.

The other information that Stanley had for us was that, while he was fishing the trout stream where it ran through the Hemlock gorge below the bridge, he had seen a water-thrush fly out from a spot along the bank and he was pretty sure there was a nest there, though he didn't stop to look because he had more important business on hand. The Artist and I went to the spot indicated and began to search both banks of the stream, he on one side and I on the other. Within fifteen minutes the Artist gave me the "come hither" sign with a waving arm while he peered at something just under the top of the bank and perhaps two feet above the swirling water of the noisy stream. I made a precarious crossing on stones slippery with dripping moss and joined the Artist. There on a little shelf in the bank was a nest fashioned out of moss, dried grass and dead leaves, and we could see four young in it.

Now the question was: Which water-thrush? The bird had flown out as the Artist came slowly along the bank on his search for the nest. He had no chance to determine its identity as it dashed downstream out of sight. We crossed the stream and climbed a bank on the far side, taking a position about forty feet away behind a hillock in the woods, so that we could peer over the rise at the nest that was below us and on the far side of the brook. At this distance from the stream we could hear other sounds than the rushing of the water, and one of the sounds we heard was the alarm note of the water-thrush. It was dodging around the trees, peering at us from different directions, but giving us no clear look at it through the foliage.

We sat quietly and waited for the bird to calm down and take up the task of bringing food to the nestlings, which were in the half-grown stage. When it approached the nest, we would have a clear view of it and the eye stripe would tell us which water-thrush it was, the Northern or the Louisiana. We sat there under the Hemlocks for twenty minutes, with mosquitoes gnawing away at us and three different birds fussing around us. One was the water-thrush that always managed to keep behind a branch or a twig or a leaf so that we couldn't get a full view of it through our glasses. Another was a Phoebe carrying food in its bill, indicating that it, too, had a family in the vicinity and was disturbed by our presence. The third was a Blue-headed Vireo that was feeding itself close by in leisurely fashion and apparently had no fear of us whatever. It came within five feet of us on a sapling, saw us plainly, heard us talking and paid not the slightest attention to us. Gradually, but not until we had to shift position several times to ease our cramped legs, the water-thrush began to approach the nest from upstream. When it alighted on a boulder in midstream I had a clear look at it and the white eye stripe stood out like a name plate; our bird was the Louisiana Water-thrush. It had food in its bill and was heading timorously for the nest and young, so we backed down out of sight in the woods, crossed the stream much higher up and left the mother and young to their family life.

We might have found the nest of the Olive-sided Flycatcher the next day except that we didn't have time to sit out the task or solve it another way by energetic inspection of all possible nesting sites in the neighborhood. It was at a lonely ice pond and we went there because we find Olive-sided Flycatchers there every year. We often hear the call note almost half a mile away as we walk toward the pond on a quiet day. This June day there was no call note and we were wondering what had happened to our expected birds as we went downhill through the pasture to the pond. We stood by the ruins of the old icehouse and heard a Canada Warbler singing in a slash on the east side of the pond. The area had been lumbered a few years earlier and was in a messy state. There were abandoned logs and disorderly piles of "lop and top" and, with new growth pushing up rapidly through such debris, it was a struggle to go ten feet in any direction.

We were pushing, stumbling and staggering through swampy ground and scrambling up and down hillocks with fallen timber and piles of brush as obstacles when we heard alarm notes that were new to us and saw two birds flitting around some thirty or forty feet over our heads. We turned our glasses on them. They were our Olive-sided Flycatchers and it was evident that we had disturbed them no little. We watched them for a few minutes and concluded that they had a family somewhere among the trees in the swampy region by the side of the ice pond, but we couldn't linger long enough to "sit them out" and track one bird or the other to the nest.

On the way home that day we nosed something decaying as we went up the wood road above the Artist's house. It was such a strong odor that we stood on the road for a minute and peered around. I picked up an empty shotgun shell on the road.

"There!" said the Artist. "Somebody shot something along here—and that's what we smell."

There had been a school at the roadside there many years ago, but only the foundation stones, a few timbers and some weather-beaten clapboards remained. One section of clapboard showed signs of having been gnawed by Porcupines.

"Maybe the fellow shot a Porcupine," said the Artist.

The next day we were walking down the same wood road and, as we approached the odorous spot, with a great flapping of wings a Turkey Vulture rose from behind the stone wall that bordered the road and beat its way frantically upward through the branches of the trees to disappear from sight. We climbed over the wall and

126

there, in the underbrush, we found the remains of the Porcupine that—as the Artist had suspected—had perished a victim to the stroller with the shotgun. It was highly malodorous, but undoubtedly the Turkey Vulture had been feeding on it with delight.

Now, a Turkey Vulture is not a common sight in the Berkshires and, though the bird seems to be on the increase in the area, it still is an event to come upon one. Furthermore, the fact that it was feeding on the putrid remains of the defunct Porcupine brought up another matter that I discussed with the Artist. Some years earlier I had sat around a luncheon table at which the late Dr. Frank M. Chapman, the then dean of ornithological experts of North America, took vigorous issue with some other luncheon guests, also naturalists, who insisted that Turkey Vultures had no sense of smell and found all their food by sight. Dr. Chapman told how he had conducted experiments at his noted station on Barro Colorado in the Canal Zone and how vultures had located decaying meat that he had hidden; hence they had to locate it by a sense of smell. One of the non-smelling party then said that Dr. Chapman must have done a bad burying job with the bait and the birds suspected the presence of food where the ground had been disturbed. Dr. Chapman insisted that there was no outward or visible sign of the hidden meat. The debate waxed warm. I kept quiet, knowing nothing about the subject matter, though complete ignorance has not always been, in my case, a restraining influence. But if a Turkey Vulture can't smell out a concealed dinner, how did this one in the Berkshires find a dead Porcupine lying just over a stone wall in the thick undergrowth of a dense wood in the full-foliaged month of June? There may be some other explanation than a sense of smell, but the Artist and I couldn't think of it at the time. I left him to ponder the problem in the Berkshires and I went back to town.

One day later in the month I was alone on the bank of the Van Cortlandt swamp. Herman the Magician was off somewhere whipping a trout stream to a froth and the Medical Student was tied down by laboratory work. The Astronomer had gone off to Alberta, Canada, to take colored photographs of the Sun Dance of the Blood Indians—and some bird pictures along the way, of course. The Red-winged Blackbirds were loud in the reeds, and every so often I would hear the ringing songs of Swamp Sparrows and the rattling songs of Long-billed Marsh Wrens. I strolled along the bank, keeping an eye out for anything out of the ordinary. I saw nothing extraordinary, but I did hear something strange. It seemed to come from the center of the swamp to the northward and, heard faintly at a distance, it sounded as though it might have been a White-eyed Vireo. But as I walked nearer and heard the bird more clearly, I knew it was a flycatcher of some kind and one that was a stranger to me.

Since it was a flycatcher call with which I was unfamiliar, it narrowed down—provided it was a native bird and not an escaped cage bird such as had puzzled us on several notable occasions—to two possibilities: it had to be either an Alder Flycatcher or an Acadian Flycatcher. I was well acquainted with the voices of all other native flycatchers that have been recorded from our region. I soon located the strange bird perched atop a small White Ash tree and, through my glasses, made certain that it was one of the little fellows of the genus *Empidonax,* of which we have four species that have been found in our territory, the two named and the Yellow-bellied and Least Flycatchers. A curious thing about this group, aside from the fact that the Acadian Flycatcher is a real rarity for most observers in our area, is that when they are vocal they can be identified blindfold as far off as they can be heard, but if the birds are silent, it takes a good view in a clear light to give the most expert field observer a clue to which of three possible species it may be. The Yellow-bellied Flycatcher may be identified by the characteristic that gives the bird its common name, but the Least Flycatcher (often called the Chebec), the Alder Flycatcher and the Acadian Flycatcher are so similar in size and plumage that, unless they are announcing themselves plainly, even the experts may be baffled.

This *Empidonax* in the White Ash tree standing out in the swamp was announcing himself plainly enough, but I didn't recognize the name. It seemed to be saying rather briskly: "Who's game? Who's game?" While I was watching the bird and puzzling over its song, along came Eugene Eisenmann, the esteemed vice-president of the Linnaean Society of New York and a fine field man on bird identification. Before he was in any position to see the bird he called out to me: "I hear an Alder Flycatcher! Is that what you're watching?"

It was the first time I had met the species in the open, and since it is always a thrill to "log a lifer"—that is, to see a bird for the first time—I was mightily stirred by this Alder Flycatcher. It stayed around the swamp for more than a week, during which period I heard it and saw it on three or four occasions. I have seen none since, but I feel confident that its notes will identify it for me the next time any Alder Flycatcher sounds off where I can hear it.

I saw the Black-billed and the Yellow-billed Cuckoos in the swamp this same day. As a matter of fact, it seemed to be a good year for cuckoos. Observant citizens will have taken notice that there are annual variations in the abundance or scarcity of various common forms of wildlife around us. Some years we find an extra-large supply of Robins, Baltimore Orioles, blackberries, butter beans, apple tree borers, vacuum cleaner salesmen and magazine vendors of middle age who knock on the door and beg subscriptions on the plea that it will help to send them through college. Other years will bring a scarcity of such forms of life and an unusual abundance of woodpeckers, clingstone peaches, potatoes, women wearing slacks, homicides and Japanese Beetles. One year we have no apples worth mentioning in the Berkshires and the next Autumn our trees produce so many that the leftovers litter the ground to the great detriment of rural morality and the milk supply. The cows munch them in large quantities and thereby grow drunk and disorderly on the ultimate cider they grind out of such treacherous fodder. Cows being ordinarily such sober and dignified creatures, the sight of them "foxed with liquor" is all the more shocking on a staid New England farm. But to return to the cuckoos, we are glad when they are in abundance because they seem to relish a hearty meal of Tent Caterpillars. We always have a superabundance of Tent Caterpillars in our territory, and it's a cheerful and encouraging sight to see some cuckoo wolfing 'em down.

9

Under the greenwood tree,
Who loves to lie with me,
And turn his merry note
Unto the sweet bird's throat?

As You Like It, Act II, Sc. 5

The shade trees, the groves and the woods of our region often are a sad sight in June because of the villainous work of vicious pests commonly called "inch-worms" or "loopers" or simply "nasty little caterpillars" by indignant citizens. Insect experts call them *Geometridae,* which is a good name for the foul family because it means "earth measurers" in Greek, and the way they loop along does make it look as though they were taking measurements. But when they stop to eat is the time they get in their deadly work. Countless millions of them will swarm over an area and leave only the bare ribs of the foliage to the aghast gaze of despondent residents. I have stood in the woods of a May morning

132

and heard their droppings pattering to the ground like a fine rain all around me.

There are different genera and a number of species, but all merit our deep disfavor. The only difference to the ordinary landowner is that some species get in the deadly work in the Spring and others in the Summer or Autumn. The pests have their cycles. Some years the trees in our territory escape with little damage and other years the confounded *Geometridae* eat everything green and leave an area looking as though it had been seared by a quick flame. I once counted fifteen of these munching caterpillars on a single leaf of a Linden tree.

I took notice at the last high tide in the affairs of these dastardly destroyers that they had a preference when they settled down to feed. Their favorite dishes were our elms and oaks. They finished off those trees first. Then they went after our maples and ashes, after which they attacked birches and dogwoods, and some even descended so low as to browse on the tough foliage of privet hedges around suburban lawns. They did not eat the leaves of our fine Tulip trees, which left me thankful but a trifle puzzled. These caterpillars do most of their destructive work in late May in our territory and then disappear almost in a body about the first week in June. They go underground a few inches and wrap themselves in cocoons of their own spinning for a sleep of some months, during which they gradually change form according to insect custom. When they emerge in the Autumn the males are winged and the females wingless moths. Since the unwinged female has to climb up the tree to lay the eggs for next Spring's hatching of pestilent caterpillars, a band of some inexpensive sticky stuff like old-fashioned axle grease around the tree trunk four or five feet above the ground provides the best barrier against such a vicious feminist movement.

I admire insects as a class and know that many of them go through life cycles that make the most adventurous human biography seem drab and monotonous by comparison, but I am opposed to the *Geometridae*— it's a family quarrel—because they do so much damage to trees. Unblushingly I admit that I love trees, and the ardent affection, dating back to childhood, only increases with the passing years. What is more wonderful to behold than the columnar trunk of a Tulip tree? There is a spot along the river road where, on my regular morning walks, I come out of the woods and see two great Tulip trees standing on a knoll with the broad expanse of the Hudson River and the Palisades as a background. The trunks reach up like Doric pillars to hold the green crowns of foliage against the sky. With the river below, it is a scene that might well have been painted by Monet, Sisley or Pissarro. I see it every morning when I am at home, and never yet has it failed to stir me afresh as a thing of beauty.

I have a brooding affection for our great Hemlocks. I dote on the dappled bark and the widespread branches of the Sycamore

> *Delaying as the tender ash delays*
> *To clothe herself when all the woods are green.*

134

Is there anything more pleasing to the eye or more soothing to the troubled mind than an American Elm on a well-kept lawn? What tree has more graceful lines or richer foliage? What of our Black Birch, with its pendulous branches and its harvest of tiny seed clusters on which Goldfinches, Siskins and Redpolls delight to dine in Winter? Purple Finches and Evening Grosbeaks love to munch the seeds of the White Ash or the Tulip trees. The Flickers and Robins that sometimes spend the Winter in our neighborhood feed on the fruit of the Staghorn Sumac and the Smooth Upland Sumac, and if we go out to look for Pine Grosbeaks of a December or January day, we make a point of looking over every sumac patch that we know because we so often find them feeding in such places.

We have a fine assortment of oaks in our territory. It is a great family of useful and ornamental trees that provide shade in Summer, acorns in the Autumn and logs for the open fire in the Winter. There are oak leaves of all shapes and sizes, some deeply cut and some with no indentations at all. There is considerable variation in the size or shape of leaves of a single species. The leaves, consequently, are not always a clear and reliable guide to the identity of any particular oak. Nor is the bark of the tree always a definite indication of the species. The acorns probably are the best guides to the identities of the different oaks. Each species produces an acorn of distinctive size and shape under normal conditions. One who has made himself familiar with acorns usually can tell at a glance what species of oak produced

it. But since individual oaks not only vary in output of bark and leaves but also have been known to interbreed, there are "problem oaks" that only a botanist can pin down. We usually name our oaks in the Autumn when we can weigh and consider at one and the same time three important pieces of evidence: the bark, the leaf and the acorn. And the greatest of these is the acorn.

Our most abundant oaks are the Pin Oak, the Black Oak, the Red Oak, the Rock Chestnut Oak, the White Oak and the Swamp White Oak. There are in Forest Park in St. Louis some wonderful Mossy-cup Oaks, huge, towering, broad-shouldered trees that I long had known and greatly admired when, in 1928, the baseball World Series brought together the New York Yankees and the St. Louis Cardinals. As a baseball writer I had been visiting St. Louis since 1922 and, on such trips, the ball clubs stayed in a hotel facing Forest Park. Every morning I wandered through the park. There are ponds and basins filled with lovely water lilies of dozens of species and colors. There is a fine zoo in the park, and there it was that I first saw the system of using moats instead of bars to keep the captive animals within bounds. There is a beautiful art museum atop a hill in the park and, in a dell just a few furlongs distant, the famous "shell" for outdoor opera and musical programs of any kind, with seats rising rank on rank in the natural amphitheater in front of the stage. There are always birds in the park, and I carried my field glasses with me when I was traveling with baseball clubs or going on other journeys to cover sports events of different kinds.

On this October day in 1928 I was prowling under Mossy-cup Oaks in Forest Park early in the morning, looking through the dried grass and dead leaves for those striking acorns that give the species its common name, a large acorn with a fringe of silvery fuzz or short silky "hair" at the rim of the cup. While I was poking the leaves and grass about with a stick I heard somebody call from a roadway nearby: "Hey! What in thunderation are you doing there?" I turned and saw that it was Judge Kenesaw Mountain Landis, the thin-faced, resonant-voiced jurist who had stepped from the Federal bench to the office of Commissioner of Baseball. He was in St. Louis as the Grand Panjandrum of the World Series and, as it was his custom to take a brisk walk every day of his life, he was taking it in Forest Park this bright October morning and caught me pottering about the park property like a man bent on gathering mushrooms. I told him that I was looking for acorns.

"Hogs and squirrels eat acorns," said the Judge energetically. "Are you keeping hogs or squirrels in your room at the hotel?"

I explained apologetically that the acorns I was hunting were a special brand, the product of the lofty oak under which he had caught me prowling, and very pretty things they were to see, too. I told him the name of the species and said that if I found any acorns I intended to take them back home with me and plant them. I had read in books that Mossy-cup Oaks grew in my home territory, but I never had run across any in the wild. My hope was to set out a few acorns that would grow to be Mossy-cup Oaks that I could see without going all the way to St. Louis.

"A likely story!" was the Judge's decision as he brandished his cane and resumed his walk.

I had found no acorns up to the time when the judicial investigation occurred, but thereafter I picked up a fair number—perhaps a dozen—of sound acorns of large size with fine fleecy fringes, and I still had them in my pocket when I went out to the ball park that afternoon to cover the World Series game. Judge Landis was in his front-row box looking out over the field where the players were warming up when I tapped him on the shoulder, reached into my pocket and hauled out four or five fine Mossy-cup acorns to display as trophies of my hunt.

"Well, I'll be damned!" said the Judge.

They really were prize specimens. I planted them behind the house when I returned home and three of the acorns sprouted. But I had no luck with the seedlings. Some boys started a brush fire that burned them back when they were three years old. The seedlings survived that ordeal by fire, but some years later, when they were about four feet high, my home was taken by the city for the construction of the Henry Hudson Parkway and we had to move out of the path of the oncoming steam shovels, graders and concrete mixers. I dug up my precious Mossy-cup Oaks and, having no spot of ground of my own at that moment, I set them down tenderly on the property of neighbors who promised them a good home as they grew up. But they did not grow up. The

139

transplanting discouraged them so much that they merely lingered until Winter came along and then they gave up the ghost.

I lost another oak sapling in that general removal, but the greater part of the story of this other oak belongs to an old gentleman for whom I had a great admiration and affection. He was Harry M. Stevens, founder of the famous catering firm that feeds so many millions of spectators a year at race tracks and major-league baseball parks. Mr. Stevens was born in Derbyshire, England. He came to this country as an energetic young man and, as was so often written about him, "ran a peanut into a million dollars." Sometimes he was dubbed "The Hot Dog King" because he was supposed to have been the originator of the "frankfurter and roll" combination (with or without mustard) for outdoor eating. He was a great character, a fine Shakespearean scholar and a lover of poetry in general. He had a fiery temper and a flair for the dramatic and he would, in denouncing somebody or something that he didn't approve, fling out savage classical quotations as effective as heaving rocks.

Old Harry was intensely proud of the United States citizenship he acquired as soon as he could, but he always retained a nostalgic affection for Merrie England and kept in constant touch with his relatives and old friends back in Derbyshire. One of those friends sent him a brace of Pheasant shot in Derbyshire. When the dead birds were delivered in New York, Mr. Stevens was just leaving for a visit to his original American home

140

in Niles, Ohio, which is where he went to work in a steel mill after his youthful arrival from England. At this Ohio home the Pheasants were cleaned for cooking and it was revealed that their crops were distended with acorns. Some of the acorns were planted. At least one of them sprouted and grew to be a tree big enough to produce acorns of its own. I was presented with an acorn from this second-generation English oak and I planted it close to my Mossy-cup group. The acorn duly sprouted and the little tree was about three feet high when the same fate fell upon it that overwhelmed the Mossy-cup younglings. But at New Rochelle, N.Y., on the lawn of Mr. J. B. Stevens, there flourishes a stalwart "third generation" tree of this remarkable line—the original oak a resident of Derbyshire, England, the second generation springing up at Niles, Ohio, from an acorn that made the transatlantic trip concealed in the crop of a dead Pheasant, and the third generation in Westchester arising from an acorn dropped by the Ohio oak. Perhaps there has been some interbreeding and the Westchester specimen is no longer of unadulterated English stock, but it would take an oak expert to determine such matters. I prefer the poetic view and look upon the ground shaded by this oak in Westchester as

> *. . . some corner of a foreign field*
> *That is forever England.*

The most romantic story of any native tree, I think, is that of the *Franklinia alatamaha*, which has no common name so far as I know. It was the Astronomer who introduced me to this tree and told me its fantastic history. Nearly two centuries ago the best botanist in North America was John Bartram, and after him came his son William, who followed in his outdoor footsteps and went even farther afield in search of native plants. Just before the beginning of the American Revolution, William Bartram traveled all over the Carolinas, Georgia and Florida, on foot and horseback, making notes of all he saw in the wildwood and bringing back a fine collection of botanical specimens for the famous Bartram nursery just outside what were then the town limits of Philadelphia. Whether or not he brought back a specimen of the *Franklinia alatamaha* is not definitely stated, but he did write of his trip in 1773 that, along the Alatamaha (now the Altamaha) River in Georgia, he "was greatly delighted at the appearance of two new beautiful shrubs in all their blooming graces. One of them appeared to be a *Gordonia,* but the flowers are larger and more fragrant than those of the *Gordonia Lascanthus* (sic), and are sessile; the seed vessel is also very different."

In his account of a later trip through the same region William Bartram wrote: "I had the opportunity of observing the new flowering shrub resembling the *Gordonia* in perfect bloom, as well as bearing ripe fruit. It is a flowering tree of the first order of beauty and fragrance of blossoms; the tree grows fifteen or twenty feet high. . . . This very curious tree was first taken notice of about ten or twelve years ago, at this place, when I attended my father (John Bartram) on a botanical ex-

142

cursion. . . . We never saw it grow any other place, nor have I since seen it growing wild in all my travels. . . ."

Nor has any other person ever seen it growing wild since that day. On either the first or second trip to that region of Georgia, however, William Bartram must have brought back specimens that flourished in the Bartram nursery and the descendants of which wild stock still flourish in cultivation around many homes as far north as Newburgh, N.Y. When William Bartram decided that his new and beautiful flowering tree was not a species of *Gordonia* because of "striking characteristics abundantly sufficient to separate it from that genus," he put it in a new genus "honoured with the name of the illustrious Dr. Benjamin Franklin" and called it *Franklinia alatamaha,* which not only "honours" Franklin but also preserves a record of the region in which the tree was discovered.

There are several curious things about the flowering habit of this tree. One is that the Bartrams found it flowering in June in the wild, whereas in cultivation it customarily begins to flower in August and continues to bloom until frost arrives. Another oddity is that one of the cream-white petals covers the others in the bud like a cap and, in the opened flower, that petal has a quaint appearance, something like a bowl with a recurved rim or a Mercury's hat minus the wings of Greek mythology. But the most curious thing of all is that we now have in cultivation in abundance a beautiful native flowering tree that was found growing in Georgia in 1773 and that no human eye has beheld in the wild state since John and William Bartram, its original discoverers, were gathered to their forefathers. There have been "lost birds" and "lost tribes," but this is the strange story of a "lost tree." The only parallel I know to this botanical oddity is found in the animal realm. Camels once existed only in the wild state. Now they exist only in the tame state and are bred and trained as beasts of burden. It's true that, under human care, the formerly wild *Franklinia alatamaha* survives in more limited numbers and over a lesser range than the camel tribe, but the tree in flower smells much sweeter than any blooming camel.

I ran into real trouble with another cultivated tree in our neighborhood. One windy and overcast day in Autumn we were walking past a fine estate when I saw some seed pods of the bean type scattered over the road. I looked up and saw that a tree growing just inside the stone wall on the right side of the road had hundreds of those seed pods still attached to its twigs. There were no leaves remaining on the tree, which was about thirty feet tall. Even the Astronomer didn't know the tree after scanning its shape, its bark and the seed pods that we picked up for scrutiny. There were many imported flowers, shrubs and trees on this estate and I volunteered to track down this strange specimen in a botanical tome that dealt with such matters. I had just such a thick volume at home and this would give me a chance to use it. Herman the Magician and the Medical Student offered to bet me on the spot that I never would track down the tree. I swore I would. The Astronomer was to be the umpire. There was to be no time limit on my

144

detective work but, while I was working on the problem, the Astronomer would send one of the seed pods to a friend and plant expert at the New York Botanical Garden for identification. The answer from that scientific source was to be kept from me until I offered my own solution.

I went to work on the supposition that, from its bean-like seed pod, the tree belonged to the *Leguminosae* family, so I lifted down my huge green volume that had to do with cultivated plants of all kinds and began reading all about *Leguminosae*. I worked my way up the trunk of that family tree and quickly passed branches such as "Edible Plants," "Forage and Cover Crops" and "Garden Flowers, Mostly Annuals." But I went out on the next branch, which was "Shrubs, Trees and Woody Vines," and found fifteen genera listed. I looked up each genus in turn and concluded from the evidence at hand —just the seed pod—that my problem tree was the *Sophora japonica*, a decorative species imported from Japan and hardy enough to survive under cultivation in our region.

At the next outdoor gathering of our strolling group I reported—but only "tentatively," I assured my listeners —that the strange tree might be a *Sophora japonica*, whereupon Herman the Magician and the Medical Student looked at one another and burst into roars of laughter. I immediately reminded them that my identification was merely tentative and I was still studying the problem. After all, the tree was bare when I went to work on it. All I had to go on was the fruit, and bean-

145

like fruits sometimes have as much resemblance as peas in a pod. I could be mistaken. I would wait for the leaves to appear in the Spring. The leaves would be a further clue; perhaps a definite one.

I did wait—for months—and the leaves were doubly compound, as described for the species. But other members of this family have doubly compound leaves too. Once again when I mentioned *Sophora japonica* as a possible solution, Herman the Magician and the Medical Student waxed hilarious.

"Is that your answer?" roared Herman provocatively.

"Come, come; make up your mind!" said the Medical Student laughingly.

Evidently I was on the verge of some ludicrous error. I decided to wait for the flowers to appear. The tree blossoms late in the Summer, so it was almost a year from the time I first picked up the fruit until I had all the evidence of leaf, flower, and fruit in hand. Then I was in despair. I went through my reference book again. I went over the flower spray with my magnifying glass. On our next trip afield I said mournfully:

"I give up on that tree. All I can make of it is a *Sophora japonica.*"

"Ho-ho!" said Herman the Magician. "That's exactly what it is!"

"We knew it long ago!" said the Medical Student with a wide grin. "The Astronomer showed us the answer from the Botanical Garden expert."

I had no gun with me or I might have shot them on the spot, and I think any fair-minded jury would have called it justifiable homicide. In any event, the whole thing cured me of what might be called "extracurricular work" or delving into the question of cultivated and imported species in our territory. After that I stuck to native products of our territory. There are enough native trees to keep me awe-stricken with admiration and faithful in my attendance on them all the year around. I like to think, with Thoreau, that the long, sharp buds of the Beech in mid-March are "the spearheads of Spring." I dote on the cunning way in which the unfolding leaf of the Tulip tree raises itself out of the bud like a miniature flag going up on a taxicab meter. I love to linger by a Linden in flower on a warm June day with the air redolent with the perfume and musical with the hum of Honey Bees at work among the blossoms. The Autumn colors of our oaks and maples, our dogwoods and viburnums, our Sweet Gum and Sour Gum, are breathlessly beautiful. Our lovely evergreens give us a measure of protection from the stark winds of Winter and offer an entrancing spectacle at each fresh fall of snow. I can understand how some primitive peoples made gods of their trees and bowed in worship before them. There certainly are moments when, gazing at a great tree, my feelings approach adoration.

10

Wherein I spake of most disastrous chances,
Of moving accidents by flood and field.

OTHELLO, ACT I, SC. 3

Through the war years, largely because of the rationing
of gasoline that kept my car anchored in the garage,
my visits to the Berkshire region were few and far between. My little cabin in the woods on the ridge north
of the Artist's stone house was untenanted for long
periods. One of my infrequent trips to the cabin, however, was fraught with danger to life. It was the only
time in my checkered career that I came close to
drowning on a steep hillside under a flood of water that
I brought upon myself.

There is running water in the cabin, but we have the
system drained in the Autumn or there would be a general bursting of pipes as soon as the mercury in the local
thermometers settled down for the Winter. I had Bill,

the plumber who put in the pipes when the cabin was erected, do the draining in the Autumn. First he turned off the inlet valve that was ensconced in a sunken box just outside the cabin. Then he packed the valve with dead leaves and old rags, padlocked the box, hung the key on a nail inside the cabin door, ran off the water that was in the cabin pipes and went his way.

One June day I arrived at the cabin and decided to turn on the water. Bill had left instructions with the Artist as to how this was to be done. He said to find the key to the inlet valve box, unlock the box, remove the anti-freeze packing of dead leaves and old rags, give the valve a full turn to the right and presto! the water would run in the cabin. With implicit trust in Bill, the key was found, the padlock removed, the lid of the box turned up on its hinges, the anti-freeze packing pulled out from the box to disclose the inlet valve and the valve given a full turn to the right. The result was a loud hiss that came with water spurting out of a petcock just below the valve.

That was just the beginning of the fun. I couldn't turn off the petcock with my fingers. It was too hard to turn by that method. I shut down the inlet valve again and went along the hillside to the Artist's domain to borrow a pair of pliers. The Artist ceased work at his easel, dug up a pair of pliers for me and then decided to go along with me in case I needed an extra hand at the water-works. The pliers closed the petcock easily enough, and once again I turned the inlet valve to the right to send water into the cabin. In a matter of a minute or so there was the cheerful gurgle of running water—under the cabin! It began to splash away in great style. I looked under the cabin through the open side of the inlet valve box leading in that direction, and my first glance in the gloom disclosed that the water was pouring out of an opening in the bottom of a horizontal pipe. Evidently a plug had been removed. I hauled myself up out of the box and turned off the inlet valve again.

"Oh, I guess Bill forgot to tell you about that," said the Artist with a chuckle. "That's the washbasin pipe. He removed the plug to make sure that all the water drained off. You'll find the plug on the shelf in the bathroom."

I went into the cabin, looked on the bathroom shelf and found—two plugs! Which was the right one? I had to go down under the cabin to settle that matter. Only a trial would tell because they were of different size. The only way to get under the cabin was to crawl through the inlet valve box. I never did understand why Bill made that box so small in the first place. He had a hundred acres of excess ground at his disposal if needed and he built a box so small that he had to wriggle through it like a large lizard when, in the course of putting in the pipes, he had to go under the cabin to wipe a joint or tighten up a coupling. Now I had to go down through the confounded box and it was a brutal squeeze. Not only that, but when I reached my destination under the cabin, I was lying on my back in a mudhole made by the water I had turned loose a little earlier.

I tried to screw one plug into the hole in the pipe just

150

an inch or two above my nose. The plug didn't fit. I tried the second one and it fitted. When I had it screwed in tight, I called to the Artist outside to reach down and turn on the inlet valve while I watched for results in the dark. He turned on the water. I heard it running and saw that the plug I had inserted was holding firm. For ten seconds I was elated, and then a loud stream of water began pouring down just a few feet away from where I was lying on my back in the dark. It was high tide under the cabin in no time at all as I shrieked to the Artist to turn off the water again or he would have a drowned body to haul out from under a building.

The second plug was the answer, of course. Bill had removed that plug from the lowest point of the water line that was the source of supply for the kitchen. I wiggled on my back underneath the cabin, pushing with heels and elbows and plentifully plastering myself with mud in the process, inserted the second plug at the right place, screwed it tight and called to the Artist to turn on the water again. With both plugs firmly in place, he could open the inlet valve with full confidence that at last everything was all right and no longer, with a flick of the wrist, was he taking a chance of losing an old friend in a drowning accident. He turned on the water. This time I could hear it surging sweetly through all the pipes in the cabin. The Artist shouted down the hole that he would go inside the cabin and test the taps to see how they were working. I could hear his footsteps just over my head. I could hear the water running cheerfully from the bathroom and kitchen taps when he turned them on.

"You can come out now," called the Artist gleefully down through the floor. "All is well."

At that precise moment a flood of water descended on my upturned face where I lay on my back in the mud under the cabin. The earlier annoyances had been mere streams. This was a Niagara Falls by comparison. The Artist heard the roar of the falling water and the louder roar of the perishing victim. He took hurried and effective steps by dashing out of the cabin and around to the box to turn off the water again. The downpour ceased. It was the boiler drainpipe in the kitchen that had been left open. The Artist located the valve just under the boiler. He closed it and this time a test of the water system proved that it was in civilized working order. But I had gone to the woods for a rest and I was a waterlogged, mud-plastered wreck when I crawled out from under the cabin with the water-works under control at last.

However, the Hermit Thrushes singing in the woods around the cabin brought balm to my soul. I always find it good to be where

Solitary the thrush,
The hermit withdrawn to himself, avoiding the settlements,
Sings by himself a song.

I was so moved by the concord of sweet sounds that I wrote a letter about it to the Dramatic Critic who, at that time, was playing the part—and adequately, too, as the drama critics always put it—of a war correspondent in Chungking, China. He would be glad of a gentle note from his native New England hills. I mentioned also that the Artist and I had fished the trout stream and gained nothing thereby except an assortment of lumps raised by the bites of myriad midges. Some months later I received a reply from Chungking in which there was this passage:

"Your letter was calculated to make me homesick if anything could. And some things can, like your tale of the Hermit Thrushes singing all day around the cabin because the shade in the woods is so deep. I really would be thrilled from stem to stern to hear a Hermit Thrush, and also by the sight of the environment in which it sings, for that is quite as exalting as the song. I am glad that the midges bit you. I hope the black flies had their pound of flesh, too. There should be some application of Emerson's Law of Compensation. A man who hears a Hermit Thrush in the June woods and writes about it to an exile in China ought to be bitten by something, preferably poisonous."

Just the same, I think he was delighted to hear about the concert around the cabin and probably read the letter over and over again in Chungking with more than a touch of nostalgia. I should have told him of the Ruffed Grouse too. They drum regularly in the woods around the cabin. And the Pileated Woodpecker, a magnificent bird, occasionally is sighted on the wing or whacking away at the branch or trunk of a tree in search of grubs.

On this June trip to the woods I gleaned some wild

153

strawberries on an open hillside and ate them with relish. The Artist urged me to come back in July when raspberries and blueberries would be ripe. He knew where there were wonderful blueberry patches that always yielded fruit of fine size and flavor. Lured by this invitation, I was back at the cabin again in the third week of July and the Artist led the way on a double mission one sunny morning. We would reach the promised blueberry land in due time, but first he wanted to explore the haunt of coot and hern at the head of a little lake below the high ground on which the best blueberry bushes grew. He knew an old wood road that led down to the swamp at the head of the lake. He thought we might come upon some interesting wading birds in the wet tangle.

An old wood road becomes overgrown very quickly when it isn't used regularly and, in light clothing and short sleeves, we soon were having a ripping time as we wrestled our way through the blackberry vines that had sprung up all over the right of way. It was fierce going, especially as we neared the head of the little lake, because a sawmill had been in operation there and the usual debris littered the area, disorderly piles of bark, discarded branches and abandoned logs. The berry thickets grew up through all that confusion underfoot, and the only thing that drove us ahead was the fact that it would have been just as bad to turn back. Since we had gone that far, we thought it best to fight our way through. We plunged ahead in hope that the worst was behind us and that we might be out of the brambles any

154

minute. The sun was shining down mercilessly as we pushed and squirmed through the thorny tangle of the slash where the sawmill had operated, and my spirits were at a low ebb when suddenly, on jumping down from the stump of a tree over which I had to struggle, I found myself surrounded by bushes bearing the most wonderful supply of Wild Red Raspberries that I ever saw in my life. I turned to the Artist, who was struggling with the tangle a few feet behind me, and announced my discovery. He joined me in wonderment and delight at the amazing spectacle. The whole enclosure was red-spotted with the abundant fruit. I thought of the hanging gardens of Babylon. The berries were large and full and most of them were so dead ripe that they came loose at a touch. Indeed, many fell off and were lost in the tangle below when we touched a branchlet. The air was full of the odor of the berries, a luscious aroma.

Izaak Walton quoted Dr. Boteler as saying, in reference to strawberries, "Doubtless God could have made a better berry, but doubtless God never did." I yield to none in my admiration for strawberries, and perhaps at another time—in June, for instance—I might stand sturdily by Dr. Boteler, but on this July day, hot and bothered and ripped and torn by the savage wrestling match with the undergrowth and overgrowth of thorny tangles in this patch of wilderness, I was overwhelmed and swept away with enthusiasm when I found myself surrounded by cascades of these wonderful Red Raspberries, so plentiful that in fifteen minutes each of us

155

had picked more than a pint. We had them for dessert at dinner that night. We poured them over a mound of ice cream, homemade, a combination that I recommend highly.

We couldn't stay in that berry patch all day. We scrambled through to the head of the lake and found no wading birds at all. Up the hillside we went to the blueberry patches, which were laden with fruit as advertised by the Artist. I told the Artist how I once had trouble over blueberries. As a boy in Dutchess County I never distinguished between blueberries and huckleberries and called them by either name when I gathered them, which I did in large quantities. But I learned better when I confused the terms in print in a newspaper column that I was writing for the *Sun* (New York). I soon was set right by an indignant blueberry booster, who bolstered his argument with the full weight of the scientific authority of Dr. E. D. Merrill, the famous and learned "Administrator of Botanical Collections of Harvard University" and head of the Arnold Arboretum.

The huckleberries on a thousand hills lay crushed under the impact of the Harvard decision. The argument started when, in describing a ramble through familiar fields, I made mention in print of passing a thicket of "High-bush Huckleberry with bell-like flowers that promised a flock of delicious berries in the near future." The line of type was hardly cool before a blueberry fancier in Vermont sprang to the attack. "Delicious," he explained, was an adjective that should be applied only to the blueberry; it was a crime to attach it to a low-class huckleberry. But he supposed that, like so many benighted heathens, I did not really know a huckleberry from a blueberry, and in that thought he would forgive me if I promised never to commit the same crime again.

I should have accepted those terms and surrendered unconditionally, but I made the mistake of insisting that the difference was inconsequential. This display of barbaric ignorance in the joint blueberry-huckleberry field caused the Vermont citizen to call on Dr. Merrill to give stern judgment, which he did. The judicial decision from Harvard was that the Vermont citizen was thoroughly in the right and I was completely in the wrong, that "calling a blueberry a huckleberry is a clear libel on the former" and, further, that "no botanist ever would call a black huckleberry a blueberry!" The exclamation point showed how deeply the Harvard botanical authority was shocked at the very thought. There were added paragraphs concerning the distinctions between *Vaccinium,* the true blueberries, and *Gaylussacia,* the huckleberries of lower repute.

Thus I learned that the blueberry is the aristocrat, fit for the palate of a gentleman, a berry to be served with the richest of cream. The huckleberry is for the peasant, the yokel, and may be dished up with a sprinkling of skimmed milk. I bowed humbly to the decision of Dr. Merrill not only because of his scientific authority but because I long had admired him as a benefactor of humanity. Those who improve fruits for human consumption or spread the planting of them to new regions

where they may nourish the population are true bene-
factors of mankind, though they are much neglected by
historians. I knew that Dr. Merrill and Dr. David Fair-
child and a few more such scientists had spent years at
such work and they were my heroes.

Incidentally, Isaac D'Israeli, father of Benjamin the
Prime Minister who spelled the family name without
the apostrophe and couldn't write nearly as well as the
old gentleman, has a chapter entitled "Introducers of
Exotic Flowers, Fruits &c." in his engaging series of
essays called *Curiosities of Literature*. He gives credit
to Sir Walter Rawleigh (that's his spelling of it) for the
introduction of potatoes and tobacco to England or
Ireland and he tells of Tradescant, who went disguised
into dangerous territory in North Africa for the purpose
of "stealing apricots into Britain." He has it that cherries
came from Cerasuntis (hence the name) and were twice
introduced into England, first by the Romans, whose
imported stock eventually died out, then again by
Henry the Eighth, who apparently had time enough for
the job between quarrels with his wives. There is also
mention of the great Nicolas Claude Fabri de Peiresc,
famous botanist who flourished in the south of France
early in the seventeenth century and who brought about
the introduction of many foreign fruits and flowers (and
at least one animal, the Angora Cat) into France,
through which step many of these importations later
reached England.

By the same token, one of my English heroes is the
gentle Gilbert White, who lived and died obscurely in

the little village of Selborne while "princes and prelates with periwigged charioteers" and major-generals and politicians filled the public eye. But it is Gilbert White who is now known around the world while most of the "great men" of the England of his time are wrapped in dusty oblivion. And there was William Cobbett, who thundered and threatened from a hundred political platforms, roaring and ranting about paper money and parsons, about wheat and "the Wen" (which was his contemptuous name for London), about taxes and Parliamentary reform, confident that his printed and shouted words on such topics were of great service to his countrymen and the important achievement of his energetic life. Yet it's the mildest part of "fierce old Cobbett" that now keeps his memory fresh and green; his evident love of Nature that is made plain in his account of the birds and animals, the trees and flowers, the hills and valleys and rivers of the English countryside in his *Rural Rides.*

It was through reading the books of W. H. Hudson that I learned of Cobbett's contribution to the natural history of England. Hudson made frequent and respectful references to passages in *Rural Rides* and eventually I acquired the work in an old two-volume edition that bore honorable scars of service. I was delighted that I found Cobbett writing after a guide told him that a certain road led to Selborne: "This put me in mind of a book, which was recommended to me but which I never saw, entitled *The History and Antiquities of Selborne* (or something of that sort) written, I think, by a parson by the name of White, brother of Mr. White, so long a bookseller in Fleet Street." Cobbett later read the book and must have liked it because, being near Selborne on another occasion, he went out of his way to visit the village to see "the Plestor," "the Hanger" and other attractions that Gilbert White had described so faithfully and so quaintly in his modest masterpiece.

I think Gilbert White of Hampshire, England, would have liked the countryside around my cabin in the woods of Hampshire County, Massachusetts. It is, as the Artist said indulgently, "a little country." There are no towering peaks near us, no great valleys, no tremendous rivers. There are little farms scattered in the little valleys through which flow little rivers. You leave one little valley, climb a goodly ridge that is well wooded, and drop down into another little valley. Trees cover most of the high ground in our region, but these are friendly woods, not awesome and overwhelming like the great forests to the north. When I first came to our hillside I found the site of the cabin covered with young trees from an inch to a foot in diameter at the butt. There were young oaks and hickories and wild apple trees, but most of the growth consisted of Sugar Maple and Black Birch. The Medical Student and I went to work with axes to clear the ground and the whole hillside was sweet with the odor of the Black Birch that fell under the ax. The wild apple trees were fiendish. They were twisted and intertwined in such a way that it was hard to decide which butt to cut first and how to get at it. They were horribly

160

scraggly and they reached out with their branches to deflect the ax or catch the clothing of the chopper in a dastardly way. It wasn't long before I was approaching one of these huddled, crouching wild apple trees as cautiously as I would have approached Briareus, the hundred-handed. But the Medical Student and I bore no grudge because we knew that they furnished food and shelter to the Ruffed Grouse of our ridge, and that more than offset any little difficulty that we might have had with the ones we were forced to clear away.

11

If I know
How, or which way, to order these affairs,
Thus thrust disorderly into my hands,
Never believe me.

RICHARD II, ACT II, SC. 2

Herman the Magician has a place in New Hampshire to which he hies himself in the fore part of July. The Astronomer and I are left to face the rigors of Summer in town. But just before he rolled away toward New Hampshire in his station wagon, Herman had one last walk with us up the railroad track on the east border of our favorite swamp. On the right of way we found the remains of a Snapping Turtle that had been run over by a train.

Herman delivered a funeral oration over the *disiecta membra* of the defunct Chelonian. It was not the customary type of funeral oration, a long laudatory lie extolling the imaginary virtues of the dear departed member of society. On the contrary, Herman began by pointing to the fragments of 'the turtle with the words "On this happy occasion . . ." and went on from there to rejoice exceedingly on the demise of this victim of a railway accident much as Quintus Horatius Flaccus burst into joyous song *(Nunc est bibendum, nunc pede libero . . .* Lib. I, 27, Carmina) when he learned that Cleopatra, the Serpent of the Nile who was poisonous to Romans and the Emperor Augustus, had died of the induced bite of an Asp. Over the shattered corpse of the Snapping Turtle, Orator Herman said in part:

"We use the term 'Skunk' to describe a man we despise, but I wish to point out to you gentlemen that a Skunk is a noble animal compared to a Snapping Turtle, and I mention the Skunk advisedly because Skunks are active assistants in the annual necessary abatement of that common nuisance, the Snapping Turtle. We have proof that the Snapping Turtle criminally gobbles down large quantities of game fish, including that most delicious of all table dainties drawn from the water, the Brook Trout or *Salvelinus fontinalis,* which, incidentally, is not a trout at all but a char, though that is beside the point at the moment. A large and hungry Snapping Turtle will pull down and eat a grown duck if it can catch the duck by the leg. Where Snapping Turtles are abundant in a marsh, the family life of the Wood Duck is violently disturbed. Now, our friend the Skunk helps in this fish-and-game drama by locating the spots where the Snapping Turtles have laid their eggs for the hatching. The Skunk digs up such eggs and eats them with gusto. Thus the Skunk, by eating young Snapping Turtles in the shell, saves fish in the streams and helps to keep duck on the water. I thank you for your kind attention and I hope we meet soon again over the remains of another Snapping Turtle."

I suggested that the Skunk, in addition to eating the eggs of Snapping Turtles, also was broad-minded enough to eat any duck eggs it might find on its travels in the same region.

"Nothing is perfect in this troubled world," said the funeral orator.

Then Herman went off to New Hampshire, leaving the swamp to the Astronomer and me for the month of July. The Astronomer has been our botanical guide on all trips and, when I am the only one with him in the field, I enjoy the luxury of private lessons. I had no real grounding in botany, no knowledge of any system by which I could identify a flower or shrub or low-lying plant that I didn't recognize at first glance. I had learned common flowers the way I had learned English, by absorbing it from my elders when I was a child. Thus I could call a fair number of common species by name, but I knew nothing of botanical orders, families and genera. The Astronomer made a valiant effort from time to time to enlighten this Stygian darkness, but I have no doubt he occasionally despaired of ever teaching me anything. He gained some amusement, however, by giving all of us field examinations at intervals. He would name a species for us and then, weeks or even months later, he would pluck a leaf, a flower or a seed pod from the same plant and hold it out for us to name. He chuckled each time he caught us in error.

Until I met the Astronomer my childish method of tracking down strange flowers was to turn the pages of a flower guide until I found the portrait of one that looked exactly like the specimen in hand. Often I was unable to find such a portrait, but the effort wasn't entirely wasted. By that method I grew to be fairly familiar with the appearance of many of the common wild flowers of our territory—and there was the exercise of lifting down the books and turning innumerable pages to boot. But from continued association with the Astronomer, I began to pick up at least a faint knowledge of the science of botany, enough to make it a little easier for me to thread my way through the mazes of the plant world and track down a strange specimen, though I still made ludicrous mistakes at odd moments.

One July day I went alone to the swamp and up the ridge to the westward and along the way I picked three different kinds of flowers for the purpose of identifying them when I reached home. There were leaves attached to all three specimens and fruit as well as flowers on two of the stalks. Exhibit A was a vinelike plant. Exhibits B and C were what ordinarily are called "weeds." I have gone afield for many years and never yet have learned where "flowers" end and "weeds" begin. I am familiar with the old definition that "a weed is a flower out of place," but who is to decide when and where a flower is out of place?

I settled down in a big easy chair at home and went

to work on my three specimens. The vinelike flower and fruit I had found growing hard by a Virginia Creeper and I suspected, from its appearance, that it was a close relative of that familiar plant. I began by looking up the Virginia Creeper. It was my impression when I started the search that the scientific name of the Virginia Creeper was *Ampelopsis quinquefolia,* but when I had thumbed my Volume II of Britton and Brown to the proper page I discovered that this authority had it *Parthenocissus quinquefolia,* which was quite a shock. However, I recovered my poise and even swaggered a bit to myself when I looked on the opposite page and found a description of my gathered specimen, Exhibit A, under the name of *Ampelopsis heterophylla.*

From the start I had recognized Exhibit B as a member of the Mint Family—or at least I was fairly sure of it because the two-lipped purple flowers were growing in clusters where the leaves branched from the stem— but I was abashed when, upon thumbing the book, I discovered that it flourished abundantly in waste places all the way from Nova Scotia to North Carolina and as far west as Utah and Montana. It was *Leonurus cardiaca,* of which the common name is Motherwort. Since it is so widespread, how had I overlooked it up to this particular day?

Exhibit C was a little white two-petaled flower that ripened into—as could be seen by glancing further down the flowering spray—hairy seed pods. I hadn't the faintest idea where to begin the search for the name of this plant. How I happened to discover that it was *Circaea lutetiana* or Enchanter's Nightshade was like winning at roulette. I just thumbed a volume of Britton and Brown at random and happened to hit a page with a lucky number.

But one sweltering July day the Astronomer determined to give me a formal lesson in tracking down a botanical specimen. I had told him of my haphazard hunting and I think it must have been my method of identifying the *Circaea lutetiana* that spurred him to a desperate effort to mend my ways. We had arrived home with a hatful of loot in the form of leaves and flowers, but he chose one flowering spray, plucked from a shrub on the fringe of the swamp, for the object lesson. We knew what it was well enough, but the Astronomer was going to show me how to trace it to its lair scientifically by the use of the "key" in Britton and Brown's *Illustrated Flora of the Northern States and Canada* (Second Edition, Revised and Enlarged).

We brought out the three hefty volumes of Britton and Brown, put them on a small table, pulled up chairs and went to work as teacher and pupil. The Astronomer turned to the "key," which is the first thing in Volume I except for a few pages of introduction. Other botanical authorities have devised different "keys," but almost any one will serve if the reader knows how to use it.

"Let's start," said the Astronomer, "with the known fact that this is a plant and belongs to the Plant Kingdom. Now, the Plant Kingdom is divided into four sub-kingdoms. The first is the Thallophyta—the algae, fungi and lichens. The second is the Bryophyta —the mosses and their allies. The third is the Pteridophyta—the ferns and their allies. The fourth is the

167

Spermatophyta—the seed-bearing plants. Where does our plant fit in that scheme?"

The specimen in hand was a flowering branchlet and, since anything that flowers usually runs to seed in time, the conclusion was that our plant belonged in the fourth or seed-bearing section.

"Right!" said the Astronomer. "Now we have it hedged in to that extent. But you see—in the key here —that the Spermatophyta are divided into two classes. One is the Gymnospermae, the naked-seed plants like the pine trees with their cones. The other is the Angiospermae, in which the seeds are enclosed in an ovary."

Again it was evident that our specimen didn't belong among the murmuring pines and the hemlocks, so it had to go into the other class, the Angiospermae.

"Now we find," said the Astronomer as he ran his finger over the key, "that the Angiospermae are divided into two sub-classes, the Monocotyledones and the Dicotyledones. You will note that it says here that most of the Monocotyledones have leaves that are parallel-veined. How about our plant?"

Most certainly our specimen had net-veined leaves like an oak and not parallel-veined leaves like most grasses and such plants as iris and cat-tail. Incidentally, in our loot of the day we had brought in leafy spray of wild Yam-root and I noticed that the leaves were parallel-veined.

"That's right," said the Astronomer. "Our Wild Yam-root is among the Monocotyledones, but this plant that we are looking up has net veins and is among the

Dicotyledones. Now we see that the Dicotyledones are divided into Series 1 and Series 2, depending upon whether they are Choripetalae or Gamopetalae, which means whether their petals are separate all the way down to the base or joined more or less somewhere along the line. What do you say about that?"

If the plant had been a criminal I had been tracking down, right there I would have gone headlong from the trail. Looking at the flower, I could have sworn that it had five separate and distinct petals.

"Look closer—all the way down," said the Astronomer.

I took another look—this time with my magnifying glass—and found that the petals were united near the base.

"So it's among the Gamopetalae," said the Astronomer. "Now we're getting warm. The next thing is to decide whether the ovary is inferior or superior. That is, whether the ovary is above or below the base of the calyx. Superior in this case? That's right. Now we get down to the stamens. If they are mostly free from the corolla——"

They were. I could see that with my magnifying glass. I wasn't trusting the naked eye after making that mistake as to whether the petals were united or not.

"Well, we see here a division," said the Astronomer, "in which the corolla is essentially polypetalous—that is, the petals divided to a considerable extent."

That was where I went wrong earlier. Now it was a guide in the right direction. Certainly our flower was "essentially polypetalous" or I wouldn't have gone off the trail some distance back.

The Astronomer nodded and then asked out of a clear sky: "Is the ovary three-celled or has it more cells?"

How would I know? I hadn't any experience in dissecting flowers to discover such intimate matters. But I brought the Astronomer a razor blade and he sliced horizontally across the middle of the ovary that was located just above the base of the pistil. Under the magnifying glass the section showed that the ovary was three-celled.

"In that case," said the Astronomer, "the key puts it among the Clethraceae and we are advised to turn to Page 666, Volume II. The key only sends us to the right family. We have to run down the genus and species from there on."

So we took Volume II, turned to Page 666, where we found that the family consisted of only one genus, which was described on Page 667. There were descriptions and illustrations of two species on Page 667 and one of them, *Clethra alnifolia* or Sweet Pepper-bush, was our plant. We had known that from the start of our formal investigation, but I had learned much in the interim. I must confess that, though I absorbed in principle the secret of tracking down any plant through the use of a "key," I frequently find myself baffled by details when I try it alone and I revert to type—I go back to my old barbaric style of catch-as-catch-can research that consists mostly of guessing where the plant might be found and turning pages until I find it or give up the chase as hopeless. I can always fall back on the Astronomer. There was only one occasion when an appeal to the Astronomer left me in the position of the biblical character of whom it was written that the last state of that man was worse than the first. It was extraordinary enough to be recounted out of season.

One cold Winter day we were walking along the west bank of the Hudson River at the foot of the Palisades and I saw some empty seed pods clinging to a dried stalk about three feet high. I had no idea what the plant was and even the Astronomer wouldn't go any further than to say that it was one of the Figwort Family. When we returned home with the seed pods he looked through Britton and Brown and shook his head. The drawings and pure reading matter didn't quite fit the specimens in hand. He put back the Britton and Brown volumes, walked out the door and went to Mexico. That's right, Mexico. He was interested in the progress of the youthful volcano, Paricutin, that had sprouted from a level cornfield in Mexico and was going great guns at that time. The Astronomer planned to inspect the volcano on foot and also take colored moving pictures of it from a plane. Two days after he left for Mexico I received a postcard as follows:

"En route to Mexico. I sent the fragment of the puzzling plant to my friend Edward J. Alexander at the N.Y. Botanical Garden and asked him to let you know what it was."

The following day there came to me a brief note from Edward J. Alexander:

171

"The plant is the Yellow False Foxglove, *Aureolaria flava.*"

I hauled down the Britton and Brown volumes to check on this *Aureolaria flava* that was a perfect stranger to me. First the index was consulted. No sign or trace of any genus named *Aureolaria*. So I looked up the English name, Foxglove. That I found in the index, with the reference "Vol. III, 206–208" for False Foxglove. It turned out that what the New York Botanical Garden expert called *Aureolaria flava* was *Dasystoma flava* in Britton and Brown. And the English name for the species in Britton and Brown was Downy False Foxglove, though that's a minor matter because English or common names for flowers, birds and trees may suffer a change any time a town line is crossed. Local option is supreme in that field.

But if there is any great virtue in scientific nomenclature, the Latin names of genera and species should agree the world around. Confused by the *Dasystoma* and *Aureolaria* offered for the same genus by conflicting authorities, I took from the shelf another book on botany, *How to Know the Wild Flowers* (New Edition with Colored Plates, 1940), by Frances Theodora Parsons. And I found that the Downy False Foxglove's professional name in that guide was *Gerardia flava!* One species with three different scientific names in three reference books. This sort of thing is enough to make a man abandon botany and take up the study of racing charts in a serious way.

By the last of July I was alone in the swamp. The Artist was painting in the Berkshires. Herman the

172

Magician was fishing in New Hampshire. The Astronomer had gone to Alberta, Canada. He is always going
somewhere either to take pictures or to give lectures.
I was sitting on the bank of the swamp of a fine sunny
morning, waiting to catch a glimpse of a certain bird.
It takes patience to see this bird. It isn't a striking
beauty like the Baltimore Oriole or the Scarlet Tanager,
but I like to see it, anyway. And I know if I sit on the
bank and look out over the swamp long enough of a
Summer day, a Least Bittern sooner or later will flap
a leisurely course over the cat-tails for half a minute
or so.

The Least Bittern breeds in the swamp, feeds along
the little stream that winds through the middle of the
marsh, and occasionally "comes up for air" when changing from one feeding spot to another. That's when it
may be seen on the fly. It doesn't fly on any set schedule,
however. A searcher has to be patient and hold still
until the Least Bittern decides to take to the air for
a short journey. I never found it tiresome to wait on the
edge of the swamp. Rather the difficulty was to drag
myself away when I knew that I must leave. Life seems
to be concentrated in a marsh as explosive material is
concentrated in a shell casing. Long-billed Marsh
Wrens, Red-winged Blackbirds and Swamp Sparrows
were sounding off all over the area of cat-tails and
alders. Barn Swallows and Rough-winged Swallows
were sweeping over the reeds and open water with
chattering cries. Painted Turtles were sunning themselves in line on logs, ready to plop into the water at
any sign of danger. There were dragonflies by the

dozen—by the score, even—whizzing up and down the slow stream and out over the cat-tails. I was idly watching them—several were brilliantly colored—when I heard something like a weak squawk and looked around just in time to see a brownish bird drop down into the marsh. I had only half a look and couldn't be sure it was the Least Bittern that I had come to see. I decided to wait for a fuller view.

A Belted Kingfisher rattled its way up the winding stream. I heard a Warbling Vireo in a Red Maple on the far side of the swamp. Below me in the water the blue Pickerel Weed was past its prime and some of the Yellow Pond Lilies had a seedy look. Bullfrogs cr-r-ronked among the lily pads. There was a swirl in the water twenty feet away as a huge Snapping Turtle—an ugly brute—came up for a breathing spell and scuttled down again when it saw me. There were butterflies fluttering around—a Tiger Swallowtail, a Silver-spotted Skipper, a Red Admiral and many smaller ones that I did not recognize. A small boy came along with a lunch box in one hand and a bird's nest in the other. The nest was empty. The boy explained that a Catbird had reared a family in it. The lunch box was full, but I could see that it wasn't going far in that condition. The boy was on the stout side and he smacked his lips as he told me what his mother had put in the box: roast beef sandwiches and chocolate cake. Had I seen any young Wood Duck? He was looking for young Wood Duck. I had no Wood Duck to report, so he wandered along, protecting his empty nest and full lunch box with equal care. I swung my gaze out over the marsh again just in time to see a Least Bittern flapping along sedately. It was well worth the wait "down in the reeds by the river."

I left my seat on the bank and began to walk northward along the railroad track. There are a thousand things in the open that baffle me, but one particular thing is annually recurrent. It came up again this July day when I heard a familiar sound on the fringe of a thicket. It was the clamor of a fledgling Cowbird crying for food from some overworked foster parent. This to me is the great mystery of birdlife. If there were a feathered Court of Domestic Relations, Cowbirds would be up before it every breeding season. They are like the European Cuckoo famous in song or story that builds no nest of its own but foists its eggs on other species for hatching and the rearing of the young intruders. Our native cuckoos are respectable and law-abiding birds. It's the Cowbird that puts over the same dastardly trick in this country. The victims are usually smaller than the Cowbird species, with the result that the young Cowbird, at hatching or shortly thereafter,

175

is much the biggest of the brood, eats more than a fair share of the food brought by the swindled parents of the true offspring in the nest and eventually gains enough strength to kick out the legitimate offspring— tossing them overside to be killed in the fall or starve on the ground below while it alone survives, waxes fat and grows to wicked maturity. It's a fantastic arrangement in Nature. What principle underlies it? What philosophy can explain it?

I have found Northern Yellow-throats, Phoebes, Yellow Warblers and Black and White Warblers feeding squealing young Cowbirds, and this time I looked around to discover what species had been the victim. It turned out to be a bird with red in its eye, but the color did not come from indignation at being thus put upon. It was merely another patient, long-suffering victim of the old swindle, this time a Red-eyed Vireo. When it came along to stuff food into the yawning cavern of the clamoring impostor that it had raised instead of its legitimate offspring, the foster parent looked to be only half as bulky as the noisy, overbearing young swindler it was feeding.

It's the second part of the life cycle of these lawless birds that I find quite remarkable. The first part—the placing of the egg in the nest of some other species— might be passed off as a labor-saving device on the part of shrewd Cowbirds. But mark what follows. Young Cowbirds are hatched and reared by Yellow Warblers or Red-eyed Vireos or brooding victims of some other species, and these are the only parents and food providers that the young Cowbirds know. But when the young Cowbirds are fully grown and able to fend for themselves, they do not associate with birds of the species that reared them. No; they go off and hobnob with the first Cowbirds they meet. Soon they have gathered in flocks for a beetle barbecue or a weed-seed buffet in a cattle pasture.

Now, one thing is sure. A young Cowbird that has been reared by a Yellow Warbler does not, upon coming of age, suddenly look in a mirror and say: "Gosh! I'm not a Yellow Warbler! I'm a Cowbird! I'll go right out and join the gang." It must learn the facts of life from some other source. It certainly isn't all done with mirrors. I have read monographs on Cowbirds—and one report of a female Cowbird so erratic that it was caught feeding a fledgling of the same species!—but I have yet to come across anything explaining this point of how the young Cowbird, on coming of legal age, recognized that it was a Cowbird and moved naturally into the correct social set. The monographs over which I pored dealt mostly with why Cowbirds laid their eggs in the nest of other species and the fiendish cleverness with which they carried out their nefarious purposes. I have my own idea as to how the young Cowbirds come to know that they are Cowbirds and not Yellow Warblers, Red-eyed Vireos or Song Sparrows, as the case may be. I believe the distinction is made on a sound basis. I think they learn by ear. I long ago discovered that much can be learned simply by listening.

12

I would fain die a dry death.

THE TEMPEST, ACT I, SC. 1

I suppose that anthropologists would classify me as a "hill man" because most of my days have been spent in rolling country and that's where I feel most at home, but as far back as I can remember I always nursed a longing for "a life on the ocean wave, a home on the rolling deep."

Who hath desired the Sea?—the sight of salt water unbounded—

I have. But when I reach the ocean, I realize that it is too deep for me, too wide, too powerful—too much for me in all directions. It awes me and overwhelms me. I would love to be a salty sailor lifting a rousing chantey in the teeth of a sou'wester that howls through the rigging of a stout sailing ship, but I haven't the stomach for it. I have been horribly seasick at most

178

embarrassing moments. Since childhood I have been a constant reader of sea stories and I dote on Joseph Conrad and William McFee among modern writers. When I was a boy I reveled in *The Sea Wolf* by Jack London and a few years later I discovered *Two Years before the Mast, Moby Dick,* and other wonderful stories of men who go down to the sea in ships. I pored over poetry about the sea with fervor. I loved the tang of Kipling's seafaring verse and I memorized—out of sheer delight in them—most of John Masefield's *Salt Water Ballads.* I have always admired shipmasters; they are great men, commanding figures, in my view. But, fearful of wind and wave, I have rarely ventured beyond my own modest depth in salt water.

The closest I ever came to a seafaring career even briefly was when I was the captain bold of a 22-foot gaff-rigged sloop of sturdy construction that I sailed on Great South Bay when my family, for three successive years, had a Summer cottage on Fire Island. I was a weekly visitor, going down on Saturday afternoon and coming up to town Monday morning. The sand and surf of the Fire Island beach are famous, and when we were in swimming the terns would come winging along the line of breakers and peer down at us as they passed close overhead—the Common Tern with its shrill cries of "Ta-a-a-rr! Ta-a-a-rr!" and the lovely Least Tern with its soft "Pip-pip-pip." Sanderlings were the regular beach birds, running in and out as the foaming water swept up the sand slope and drained back again with a rippling swish. On the level part of the beach that

179

reached back to the base of the dunes we could always find Piping Plovers uttering their plaintive calls. On one great occasion the Medical Student and I, returning from an August walk of some miles along the beach, saw a rather large bird standing at the foot of the dune wall. We swung our glasses on it and discovered that it was a Hudsonian Curlew, a life record for us. The Spotted Sandpiper was a regular beachcomber and nested in the grass on the dunes.

We used our sailboat for bird hunting. From one of the marshy islands in the bay there projected a sandspit that was surrounded by mud flats at low tide, and birds of all kinds gathered there either to sun themselves on the sandspit or feed on the mud flats. We used to sail down that way—the Medical Student was my regular shipmate on such expeditions—and inspect the premises. We saw many birds during the voyage, of course, because gulls, terns and other birds were to be found on the wing or water or perched on the nets of the fishermen or the "fish traps" at the end of each line of nets. "Fish traps" of this kind jutted up from the waters of the bay in all directions and we could sail close by Laughing Gulls, Herring Gulls, perhaps a Ring-billed Gull and even an occasional Great Black-backed Gull perched on the upright posts from which the encircling nets were draped to keep the fish impounded. In September the Double-crested Cormorants came to add themselves to the seating list at the "fish traps." The terns were content with the uprights on which the nets were stretched across the bay, and there we would see

180

them—mostly Common Terns—all in a row, one bird to an upright.

When we came near the sandspit we would sail in as close as we could, throw the sailboat up into the wind, toss over the anchor, drop the sails, lash everything down and then jump overboard in our bathing trunks to wade ashore with our field glasses held high overhead. The usual occupants of the sandspit and adjacent mud flats were the Semipalmated Plover, Piping Plover, Spotted Sandpiper and the "two little peeps," the Semipalmated Sandpiper and the Least Sandpiper, which I have great difficulty in telling apart. I look, as directed by the best authorities, for the lighter color of the legs and the thinner bill of the Least Sandpiper, and I am told that the Semipalmated Sandpiper is the more common species on the ocean beaches, whereas the Least Sandpiper prefers the mud flats of nearby bays, but the birds do mingle frequently and I find it difficult to persuade two individuals of the different species to stand side by side for leisurely comparison in a good light; *hinc illae lacrimae!*

Other birds that we saw there between July and October were the Ruddy Turnstone, the Greater Yellowlegs, the Western Willet, the Killdeer and the Blackbellied Plover. We could always find the Sharp-tailed Sparrow (which I jocularly call "the Blackburnian Sparrow" for the rich color of its cheek stripes) in the marsh grass behind the sandspit and we assumed that there might be more birds than we knew in the interior of the marsh. But, as we were wearing only bathing

181

trunks, we never would have lived to tell the tale if we had gone into the marsh. We would have been eaten alive by mosquitoes. The flies on the sandspit were bad enough without looking for more trouble. I remember venturing into a marsh further down Long Island in July, and the swarm of insects I stirred up around me was so thick that I could hardly breathe or keep my eyes open with safety. I fled from the scene with insects in my eyes, nose and mouth and have kept clear of such places ever since. I will make considerable sacrifices to see birds in their natural haunts, but I draw the line at inhaling insects in suffocating quantities.

The land birds that we saw on Fire Island were comparatively few in species but numerous individually. One season the Medical Student and I counted fifty-two species of land birds on the island and twenty-two species of gulls, terns, waders, shore birds and other such followers of the water. Around the houses of our community there flourished a fine growth of Poison Ivy, every patch of which seemed to be well filled with Song Sparrows, Northern Yellow-throats, Red-eyed Towhees, Catbirds and Brown Thrashers. Barn Swallows were sweeping the air for insects through most of the day-light hours, and flocks of Cedar Waxwings were shifting around the shrubbery at frequent intervals. Kingbirds, House Wrens, Robins and Northern Flickers were seen from time to time and Starlings were always strolling, feeding and fighting on the town dump. We heard and saw Goldfinches occasionally and now and then an Eastern Belted Kingfisher would be found on patrol along the bay front. There were Red-winged Blackbirds in quantities in the marshes of the island and in the twilight there were generally a few Chimney Swifts whirling in chattering circles over our cottage.

But the best time for birds was in September when the land-bird migration was in full swing and they would come, large and small, pouring over the dunes from the eastward to drift past us and disappear to the westward. There were days when the Pitch Pines— the most abundant native tree in our section of the island—were fairly swarming with Red-breasted Nuthatches and warblers of four or five species. I believe the Yellow Warbler bred in our neighborhood, but on migration the most plentiful species were Black-throated Green Warblers, Black and White Warblers and Myrtle Warblers. We also saw a fair number of Black-throated Blue Warblers, Prairie Warblers and Redstarts. The Red-eyed Vireo was the only one of that family that came under our observation on Fire Island, and the only time we saw Bobolinks was when we heard the "clink-clink" of their call notes and swung our field glasses up to catch sight of them overhead on their southward migration in late August or early September.

Nighthawks, looking like giant swallows, passed high overhead in swirling groups and once, as the sun went down on a lovely September day, a Whip-poor-will swooped suddenly and silently like a swift shadow over the roof of our cottage and quickly disappeared in the dusk on its westward flight. But the hawks gave us our greatest thrills. The Medical Student and I would seat

183

ourselves in a patch of shrubbery on the moor between the bay and the ocean and the hawks would come streaming over the dunes from the eastward, some of them hunting so intently that they almost flew in our faces. There were Marsh Hawks, Duck Hawks, Sparrow Hawks, Broad-winged Hawks, Cooper's Hawks and Sharp-shinned Hawks. We suspected that several of the smaller falcons we saw were Pigeon Hawks but we could not be sure. One year there seemed to be an extraordinary number of Yellow-billed Cuckoos slipping quietly through the Pitch Pines and assorted shrubbery on migration, but in three seasons I saw only one Black-billed Cuckoo on the island and the Medical Student saw only two.

Most of the summer residents contented themselves with surf bathing or sunning themselves on the beach a large part of the day, but the Medical Student and I put in many hours sailing on the bay or tramping the island in search of birds. We loved to sail and we took pride in the way we solved the nautical problem of getting our boat in and out of the basin through a narrow channel with the wind coming from different points of the compass. It was fun to sail close-hauled in a brisk breeze with the lee rail down and the spray coming over the bow to give us a splashing. It was luxurious to loll in the little cockpit as we loafed along with the boom wide to a following wind, taking the little lift and heave as if we were really "rolling down to Rio." In short, our small boat was great fun.

Perhaps our trips on solid ground lacked the romance

184

of our sailing expeditions, but the Fire Island terrain was, in a way, foreign territory to the Medical Student and me and we explored it with enthusiasm. On the moor near us there were wonderful clumps of High-bush Blueberries into which we wandered to gather many quarts of fine large berries to furnish forth our breakfast tables. In some sections the open ground between the blueberry patches was almost matted with the creeping foliage of the Bearberry vines, the red berries making a pretty picture with the green leaves. There are thickets of Shad-bush, Sassafras, Red Cedar, Pitch Pine and other somewhat stunted trees, and my first venture into one of these tangles taught me a sharp lesson. I had been swimming in the surf and, when I left the beach to stroll in the hinterland, I merely picked up my field glasses and went my way. Since Fire Island is simply a wide sand bar about twenty-five miles long, it's safe to walk barefoot almost anywhere on the island. I wandered into a breast-high thicket and found myself involved in a tangle that left me with a wrestling match on my hands. The question was whether to plunge hopefully ahead or to turn back and fight my way clear again. I decided to plunge ahead. Suddenly I felt as though I were walking on broken glass with my bare feet. I had blundered into a patch of American Holly, and the sharp prickles of the fallen leaves that covered the ground were putting me to torture.

Possibly if I had been in close pursuit of some rare bird at the time I might have trod the thorny region without flinching, etherized by rapt attention on some other matter. My first experience with shore birds came some years earlier when I visited friends who had a cottage on Westhampton Beach. It's really a continuation of the Fire Island beach, but it was all a new world to me in those days and I knew nothing of such matters. My friend took me in a rowboat on the bay and, clad in bathing trunks, I left the boat in shallow water and waded ashore in order to inspect some birds on a mud flat. I was barefooted and up to my ankles in ooze near the fringe of the marsh grass when I spotted what was then a new bird for me, the Eastern Dowitcher. I moved up on it cautiously, holding the bird all the while in focus in my field glasses so that I could note all the details of its markings. When the bird finally flew off, because I had come too close for comfort, I waded out to the rowboat, climbed in and sat in the stern as my friend pulled toward home. Suddenly he stopped rowing and accused me of filling his rowboat with blood. I looked down and saw that I was bleeding plentifully from a cut on the big toe of my right foot. It was only then that I remembered feeling a pain while I was walking up on the Dowitcher, but I paid no attention to the pain in the thrill of putting my field glasses on a "lifer," a new bird for me. I was still dazed with enthusiasm when I climbed into the boat and it wasn't until the owner called my attention to the mess I was making of his boat that I realized what had happened. Evidently I had stepped on a sharp clamshell turned edgewise in the mud while I was stalking my prize bird and the edge of the clamshell had cut my toe to the bone. I would have howled aloud at the moment of incision if I hadn't been "etherized" by my rapt attention on the Dowitcher. Indeed, I did howl heartily when the doctor poured alcohol into the cut before he bound it up. It was a full week before I could wear a regular shoe again or walk solidly on the foot, but I took it in stride as an interesting demonstration of the power of mind over matter.

World War II put a stop to our Summer visits to Fire Island. The ocean front, with German submarines lurking offshore, was a guarded zone in which a bird student with a pair of field glasses would have been grabbed as a Nazi spy in a hurry. All the rollicking youngsters of our community eventually were drawn into various branches of the armed service and—some of them, at least—

> *. . . knew the misery of the soaking trench,*
> *The freezing in the rigging, the despair*
> *In the revolting second of the wrench*
> *When the blind soul is flung upon the air.*

So Fire Island became a Lost Atlantis to me, but I look back on delightful days there—days of adventure, days of discovery, days of deep elation in the sun and the surf, days filled with the fun of sailing—and nights of surpassing loveliness when, from the dunes, we gazed at a golden moon above the restless ocean and watched the grumbling rollers splash into a silver carpet as they broke on the sand below us.

13

*Here, here, here's an excellent place, here
we may see most bravely: I'll tell you them
all by their names as they pass by.*

TROILUS AND CRESSIDA, ACT I, SC. 2

We had distinguished additions to our strolling group on
some occasions and we are inclined to boast of it in a
mild way. We look back on their visits with pride and
look forward to other days in the field with such learned
scientists and friendly gentlemen. The company of one
such kindly expert, however, we shall not enjoy again.

It was through Edwin Way Teale, author of enchant-
ing chapters on Nature and maker of amazing photographs of insect life, that we met the late Dr. William T.
Davis of Staten Island residence and national fame in
entomology. On June 16, 1927, Dr. Davis had watched
what he called "Brood Number One" of the "Seventeen-
year Locusts"—this insect is really a Cicada, scientific
name: *Magicicada septendecim*—leaving their outgrown

188

shells and taking flight in a region called the Half Hollow Hills on Long Island. He had promised himself that he would return to that spot seventeen years later to watch the next generation of the brood, after sixteen years spent underground, take wing like its forebears and enjoy briefly a place in the sun. Dr. Davis was back among the Half Hollow Hills on June 16, 1944, and through the invitation and road directions of our mutual friend, Edwin Way Teale, we were there with him.

Dr. Davis was by far the oldest, smallest and liveliest person in the party. He wore a business suit and a stiff straw hat of the kind irreverently called a "skimmer." He peered through spectacles and he flourished a walking stick blithely as he sauntered along with a little knapsack on his back. We discovered that in the knapsack he carried many items that a naturalist might want in the field: a knife, tweezers, three brass-covered magnifying glasses of different powers, an assortment of small boxes to contain any specimens he collected, several packages of sweet crackers and two bars of chocolate. He was an old hand at the game and prepared for any emergency. Mr. Teale had the old gentleman in his car when we met by appointment this June day on Long Island and, when we had joined forces, we drove down a country road together until Dr. Davis, who had been on the lookout, ordered us to pull over to the roadside and park the cars where a leafy lane branched off the paved road.

Our elderly guide thought that the lane would lead us to our insect rendezvous among the Half Hollow Hills,

189

but there had been some changes in the local landscape in the seventeen years since his last visit—real estate developers had been at work—and he wasn't quite sure that he had the right lane. He walked to a house that was set back on the other side of the road and tapped on the front door with his stick. This brought no response from within the house because there was nobody at home, but two dogs rushed around from the rear to bark furiously at the trespasser. Dr. Davis dispersed the dogs with some threatening gestures of his stick and waved to us to come over and join him. There was a tree on the lawn in front of the house and Dr. Davis, pointing up the tree with his stick, said:

"Listen! That's a Seventeen-year Locust singing!"

We gave ear and heard it. The sound is difficult to describe. It's more like a thin high moan than a rattling strident buzz such as that given off by the ordinary annual "locust" or Cicada (*Tibicen sayi*) of our territory. We listened respectfully to this lone performer and then went up the lane—it turned out to be the right lane—until we came to a field where Dr. Davis, seventeen years earlier, had watched the parents of the 1944 brood clambering out of their "nymph cases" and flying off in the June sunshine. We entered the field and found two nice calves tethered to stakes. We made friends with the calves and then began to prowl the place for *Magicicada septendecim*. Dr. Davis vowed that the date was right, the place was right, and the weather was perfect for the occasion, a warm sunny day with a light breeze.

We looked high and low through the shrubbery and

190

some scattered apple trees in the field, but never a Seventeen-year Locust did we see or hear. We did find, however, plenty of signs of where their caravans had rested. Probably through favorable weather they had their change of life a trifle ahead of schedule because we found many empty "nymph cases" clinging with phantom claws to the branches and trunks of the apple trees, each case with a thin slit in the back through which the tenant had escaped to take flight in the sunlight after sixteen years of darkness in an underground home. We collected about sixty such empty cases and then Dr. Davis, flourishing his stick cheerfully, said:

"The birds have flown, but we'll get them next time. We have a date in this field for June 16, 1961, to see the next brood take off."

But the gallant old gentleman will not be on hand to keep the engagement because he died about a year after our trip to the Half Hollow Hills in 1944. He was a fine scientist and a most amiable character whose friendly spirit and enthusiasm for Natural History led many youngsters to follow him afield on Staten Island and elsewhere. Some of his students became distinguished scientists in their own right. One of "his boys" was Dr. James P. Chapin, Associate Curator of Birds at the American Museum of Natural History. Once we were lucky enough to have Dr. Chapin join us for a saunter around our home territory in early Autumn. We knew something about his work and his travels. We knew that he had been on scientific expeditions in many parts of the world and had seen almost everything there

191

was to see, high and low, ashore or afloat. He had climbed mountains and pushed his way through jungles. He had spent years in Africa and was the court of last resort in scientific decision on the haunts, habits and subspecific differences of African birds. We were delighted to have him with us on a walk. But we were cautious, too. Herman the Magician said to me in a whisper as the Astronomer and Dr. Chapin moved off ahead of us toward the swamp:

"We're over our heads with these chaps. Let's keep quiet and learn something to our advantage."

Herman was as right as rain, of course, and we listened respectfully and eagerly as the two licensed scientists discussed the identification, ecology, economy and life cycles of what we found in the way of plants, animals, birds and insects on this trip. It was a fine sparkling day and all went well until we left the swamp and moved westward to the ridge overlooking the Hudson River. There on an estate flourished the *Sophora japonica,* the exotic tree that had given me so much trouble— and the Medical Student and Herman the Magician so much merriment—when I set about identifying it. As we came under this tree—it grew behind a low stone wall but the branches stretched out over the road—I couldn't resist the temptation to make my first speech of the day, and a grievous mistake it was. I pointed to the tree and said to Dr. Chapin:

"Bet you can't guess what this tree is."

Dr. Chapin admitted as much with complete unconcern. I named it with a lofty air and waited for Dr. Chapin to bow low before me. Instead of that, he smiled gently and remarked that he had seen trees of the *Sophora* genus—species probably *toromiro*—on Easter Island when he was there in 1934. He added casually that Easter Island was so called because it was discovered on Easter Sunday, 1772, by a Dutch mariner, Admiral Roggeveen; that it probably was the eastern limit of the Polynesian migration at sea; that the *Sophora* was one of the few trees of any kind on the island; that possibly this lack of lumber for building boats was why the ancient Polynesians, great travelers by water, had not pushed on eastward from Easter Island to the western coast of South America; and that Asa Gray listed one species of *Sophora* as a native of our own Southwest. I certainly learned a lot in a hurry when I tried to baffle Dr. Chapin with an exotic tree.

I had a setback of a different sort from another famous gentleman who joined us on a trip to the swamp. This was—during World War II—Major George Miksch Sutton, author, artist, explorer and naturalist, who, at that time, was involved in tests of arctic and tropical food and clothing for Air Force personnel who might be shot down in strange places. He also took part in experimental "rescues" by helicopter of men on ice floes in the Arctic Ocean or in steaming jungles close to the Equator. He was in town on an official mission and, with a few hours to himself, he came out to join us on one of our regular rambles.

Here I have to break down to the extent of admitting that I am generously equipped with ears and, to make the best of a bad business architecturally, I am inclined to be boastful of my hearing powers. One reason why

193

the Astronomer, the Artist, Herman the Magician and other kind friends have tolerated me as a companion on trips afield is that I help them to locate rare as well as common birds in the field at all seasons of the year through the efficiency of the listening devices I always carry at the alert position, to wit: the aforesaid generous pair of ears set at a wide angle from my sconce to catch sounds of high frequency and low decibel rating. Before they could hear anything of note, I would sound a warning cry of "Hark! Hark! the lark!" or whatever it might be—Scarlet Tanager, Rose-breasted Grosbeak or White-eyed Vireo. At my warning they would "stand at gaze like Joshua's moon in Ajalon" and soon sight the bird that I had picked up first by ear. Thus, over a period of years, I had developed a reputation in our set. It vanished in one morning when Major Sutton came along.

It was in August and songs are few and far between in that month, but I knew the call notes of many of our Summer resident birds. That wasn't good enough when Major Sutton went to work. As we skidded down a bank through the Joe-Pye Weed and began to pick our way through the alders of the swamp (I think our abundant species is *Alnus rugosa*), Major Sutton identified all the birds in the vicinity simply by hearing their chips and chirps. If any lurking or passing bird let as much as a peep out of it, the Major named the species instantly. I sprained my ears badly trying to keep pace with the confounded chap in the soldier suit and soon the Astronomer was looking at me as I had looked at a pair of old shoes two days earlier. I saw that the shoes, though they

194

had served me long and faithfully, were worn through, finished off, fit for no further use. I discarded them sorrowfully. I could see that same look in the Astronomer's eye as he gazed at me in the light of Major Sutton's exploits by ear and I said to myself, borrowing from Cardinal Wolsey:

Farewell, a long farewell, to all my greatness!

Worse than that, it turned out that Major Sutton, in addition to being able to identify all sorts of pips, chips, lisps, peeps and cheeps from a thicket or on the wing, was also a Past Grand Master in the art of "squeaking" to draw birds down out of trees or up out of bulrushes into clear sight. He did it by holding his closed fist to his mouth and drawing in his breath sharply. By this method he squeaked a Yellow-bellied Flycatcher out of the alders and he squeaked a female Rose-breasted Grosbeak out of a White Ash tree. He squeaked into sight four or five kinds of warblers without drawing a long breath. By this time I was green with envy and blue with humiliation, so I led the group from thicket to thicket with water underfoot and Catbrier, Button-bush and Climbing Boneset adding to the confusion of intertwined alders and willows. I plunged ahead wherever the prospect was horrible. We sank in the marsh. Our faces and hands and clothes were scratched by the sharp thorns of the Catbrier. We were dripping with perspiration.

In the thick of the marsh I turned to survey Major Sutton and found that he was soaking wet to his knees and his uniform was all askew from the treatment it had received in the tangles through which I had led the party with malice prepense. When he looked at his wrist watch and said that it was time for him to report back for duty, I took the group out of the swamp by way of a monstrous bower of Great Ragweed in full bloom, whereby the Major acquired a coating of golden pollen to go over his bedraggled uniform. He had ruined my reputation, to be sure, but my vindictive revenge had left him with a pressing problem on his hands. He was on an official journey and that was the only uniform he had with him. He looked more like an ill-used prisoner of war than a well-dressed officer of the Air Force when he came out of the swamp with tokens of my displeasure marking him from head to foot. I hope it taught him a lesson.

I was sorry that on the cold day when we had a really big bird man with us—Dr. Robert Cushman Murphy, Chairman of the Department of Birds at the American Museum of Natural History, who is well over six feet in the perpendicular—we didn't turn up much of interest in the swamp, but a friendly hail from another strolling group gave us the information that there was a Ruddy Duck on the lake nearby. We scurried off to the lake and it was Dr. Murphy who, at a quick glance, announced that the Ruddy Duck, a female, was the one on the extreme left of a line of seven ducks on the water. The others were Baldpates, four males and two females. It was the first time I had seen the Ruddy Duck in the wild.

Edwin Way Teale joined us on occasions and we were delighted to have with us a man who knew so much about insects. One day in late August he set out with us to see what land birds were already on the southward migration through our territory, but we cut in on him with queries on insect life. Every time Mr. Teale began to focus his field glasses on a bird or bent over to have a look at a flower, one of us would shove an insect under his nose and demand its name, home address, general business activities and life history. He was watching a dozen or so Bobolinks in their yellow-brown Autumn traveling costumes swing down into the cat-tails of the swamp when the Medical Student nudged him and dropped on the palm of his hand some insignificant slug, larva, or caterpillar that he had gleaned from a bit of shrubbery.

"What is it?" asked the Medical Student.

"I don't know," said the insect expert with a shrug of his shoulders. "I'd have to look it up."

A few minutes later, when Mr. Teale was slipping around a tree trunk to watch a dazzling male Wood Duck that was swimming about the lazy stream in the center of the swamp, I found an inane-looking cater-pillar slithering along a leaf and presented it to him for identification.

"Just a caterpillar," he said coldly. "Might be any of two hundred species."

He went off in chase of a small bird that was skulking low in the alders. It turned out to be a Swamp Sparrow. Then he investigated a slight movement in the cat-tails and found it to be a Long-billed Marsh Wren. A wasp alighted close at hand and we heckled him about it.

"Genus *Polistes,*" he said, and gave an exhibition that almost unnerved us. He stuck his finger in the wasp's face—practically gave it a poke on the nose. The wasp clung to his finger as we looked on in horror.

"Notice its white face?" said Mr. Teale coolly. "All those with white faces are males and have no stings."

A man must be sure of his facts before he tries stunts like that. We were much impressed and didn't bother the insect expert for fully fifteen minutes. Indeed, it was the long-suffering entomologist who brought up the matter of insect study a little further along the trail. We were walking up a dirt path through a grove when Mr. Teale stopped and said as he pointed downward:

"Here's an Ant-lion's excavation. Let's dig him out."

We looked where he was pointing and saw, on the hard ground, a little funnel-shaped pit about an inch across and a half inch deep. Mr. Teale took a twig and began to pry into the bottom of the pit. In a few seconds he reached down and picked up an insect, saying as he did so:

"Here's the old boy now."

It was a brownish-gray insect with what looked like black incurved horns but were really sharp grasping jaws. It's with these pincer-like jaws that the Ant-lion grabs the ants that tumble into its pit. After we had looked over this curious insect, Mr. Teale put it back on the ground and said:

"Watch it dig itself backward into the ground and then start turning to make its pit."

Just as the insect expert had predicted, the amazing Ant-lion began to disappear by the stern. In half a minute it was out of sight and we could see the pit begin to take shape through the subterranean turning movements of the pit builder. Some yards ahead on the same path Mr. Teale discovered another Ant-lion's pit and, when we bent over to inspect it, we saw that there was an unfortunate ant at the bottom in the jaws of the owner and operator of the death trap. And to think that such an ugly larva as the Ant-lion turns into a beautiful flying insect with lacy wings! Truly the insect world makes the human world seem dull and stodgy.

We bored Mr. Teale with tales of the biggest thing in the insect line in our territory, the Cicada-killers that look like grandfather wasps and might be put on the game-bird list if they were any larger. They are sometimes called "Digger Wasps" because they have underground burrows in which they place their eggs and the food supply—the Cicadas they kill—for the larvae when they hatch. Our colony—it's something like a rabbit warren—is located between the ties of a siding on the railroad track that borders the swamp. The burrows run under the ties for the most part, though I have seen burrows in pastures and orchards that were simply holes in the ground with no rock or timber as partial protection or support.

The Cicada-killers are most noticeable in the Summer when they are flying around in search of their prey. They are big enough to frighten the ordinary citizen, but Mr. Teale comforted us with the information that they are not given to attacking innocent taxpayers but concentrate on Cicadas—called "Locusts" by all except entomologists—that they stalk by sound. When they hear a Cicada give off the rattling buzz that constitutes its "song," they go after it. Since the victims are about the same size as the killers, it's interesting to watch the big wasp handle the problem of transporting the body to the burying ground and pack it down the hole. The Cicada-killer lays an egg on the leg of its victim, and the egg, on the word of Mr. Teale, hatches in three days. The grub feeds on the body of the Cicada for about ten days and then spins a cocoon of silk and earth in which it spends the Winter, emerging as an adult wasp the following Spring.

The story of the Cicada-killer carried us to higher ground where I asked Mr. Teale to identify some odd insects that were sitting lengthwise on the twigs of a Wafer-ash sapling.

"These are Tree-hoppers, sometimes called Brownie Bugs," said Mr. Teale. "This species is *Echenopa binotata.*"

We all thanked him because we had seen the insects on that same tree year after year and we never saw them anywhere else in the neighborhood. It was a puzzle to us and we were grateful that Mr. Teale had named them. We allowed him a few minutes to enjoy the birds and flowers around us, and then the Medical Student came up with a little green crawling thing and asked:

"What species, please?"

"You flatter me," said the entomologist wearily. "It's

a caterpillar. Beyond that, it might be anything. Let me remind you gentlemen that there are 625,000 species of insects known to science and I don't carry them all in my head."

That ended our first lesson. Of course, we never have stopped bothering Mr. Teale about insects. We do it *viva voce* and by correspondence. I wrote him proudly that a lady had brought to the house what she said was a "bug" in a little box and I identified it for her. She said it had "great big eyes like the wolf in *Little Red Riding-hood* and it bounced off its back with a click," whereby I suspected that it was one of the "Click Beetles" and, through a reference book, identified it as the Eyed Elator (*Alaus oculatus*), so named because of black spots ringed with white to look like comparatively gigantic eyes up forward, the true eyes being much smaller and less noticeable.

Mr. Teale gave me a passing mark on that test, but more often I gave him cause for hearty laughter at my mistakes in insect identification. I humbly admit my vast ignorance in that field. Insects are marvelous, but there are too many of them. It isn't a difficult matter to track down the species of any mammal, bird, tree or flower that may be found in our territory, but the insects are overwhelming. At least, we find them so. We stop to admire beautiful insects when we see them on our walks but, except for a few common species that everybody can name, if we want to know more about them we take the matter up with some insect expert—like Edwin Way Teale.

14

COSTARD
Well, if ever I do see the merry days
of desolation that I have seen, some shall see——
MOTH
What shall some see?
COSTARD
Nay, nothing, Master Moth, but what they look upon.
LOVE'S LABOUR'S LOST, ACT I, SC. 2

We know perfectly well that all our butterflies and
moths—the *Lepidoptera* of the scientists—are insects by
legal definition, but we blandly ignore the law and, in
our own minds, put at least the larger and more beautiful butterflies and moths in a class by themselves. Not
that we know much more about them than we do of
their general classmates, but that we notice them more
often on our walks, and always with delight.

We have found the lovely Mourning-cloak coming
out of hibernation as early as March and, in the warmer

months, we are flattered to have the Monarch dancing attendance upon us as we look out over the swamp or pick our way through a meadow knee-deep in Red Clover. We often see the Tiger Swallowtail, the Spice-bush Swallowtail, the Great Spangled Fritillary, the Red Admiral, the Buckeye, the Viceroy and the Little Wood-satyr. The Common Blue and the Cabbage Butterfly travel the roads with us all Summer and the Silver-spotted Skipper bobs up out of the weeds wherever we go. We see some of the Angle-wings occasionally and once, in the late afternoon on Fire Island, the Medical Student gently lifted a wonderful Cecropia Moth from the telegraph pole to which he found it clinging.

While gathering blueberries on the moor near our cottage on Fire Island I saw a small moth so beautiful that I decided to try to identify it. I didn't have my large reference books on the island but I had a few pocket guides and one was the *Field Book of Insects* by Dr. Frank E. Lutz, which had twenty-odd colored plates in it. To my good fortune, my moth was pictured on one of the colored plates, probably because it was so attractive to the eye, and thus I learned that my flutter-ing friend of the Fire Island moor was *Utetheisa bella*. Later I often saw others of the same species in my walks over the moor. On another occasion when I was at home I found a moth of about the same size and just as bril-liantly marked fluttering about my bedroom. I caught it and found that it was a colorful moth of a kind I never had seen before. I hauled down my big Holland—*The Moth Book*, by W. J. Holland, 479 pages, "Illustrated with 1494 color photographs and 263 drawings"—and studied the plates until I found a pictured moth that matched the one I had in hand. According to the colored plate, that made my moth the *Atteva aurea,* but when I turned to the pure reading matter about that species on Page 424, I found it described as "distributed from the Gulf States southward and westward into Mexico and lands still further south." This shook my confidence in my identification, so I packed the little moth off to the Astronomer and asked him to show it to some expert at the American Museum of Natural History. There my identification was confirmed and the Astronomer was informed that other individuals of the same species had been collected in our region that same Summer and sent to the museum for identification.

I always looked upon the late Dr. W. J. Holland as the last word in authority on *Lepidoptera* and I have *The Butterfly Book* by W. J. Holland ranged with *The Moth Book* by the same author on my library shelf, but in my younger days I had only a small oblong book with a limp leather (imitation) cover, a field manual that cost $1.25 and fitted nicely in a coat pocket. In my early endeavors to identify the butterflies I found afield I thumbed this little book backward and forward hundreds of times until I had it dog-eared and ragged. I read the text over and over again and there was one page that fascinated me. Beside a color plate of Hunter's Butterfly the reading matter ran as follows:

We all know Hunter's Butterfly. How many know that its name commemorates that of a most remarkable American,

John Dunn Hunter? Captured by the Indians in his infancy, he never knew who his parents were. He was brought up among the savages. Because of his prowess in the chase they called him "The Hunter." Later in life he took the name of John Dunn, a man who had been kind to him. He grew up as an Indian, but after he had taken his first scalp he forsook the red men, no longer able to join them in their bloody schemes. He went to Europe, amassed a competence, became the friend of artists, men of letters and scientists. He was a prime favorite with the English nobility and with the King of England. He interested himself in securing natural history collections from America for certain of his acquaintances, and Fabricius named the beautiful insect shown on our plate in his honor. His *Memoirs of Captivity among the Indians* is well worth reading. In that charming book, *Coke of Norfolk and His Friends,* which recently has been published, there are some most interesting reminiscences of this American gentleman, for gentleman he was, although reared by savages. The presumption is established that his unknown progenitors were gentlefolk. "Blood will tell."

This struck me as an entrancing story and I felt I would like to know more about the adventures and striking career of such an astonishing character as the John Dunn Hunter thus brought to my attention in a book about butterflies. But it was years before I actually carried out any further investigation of the matter. One day when I was looking up a book in the Reference Room of the great Public Library at Fifth Avenue and Forty-second Street in New York City I suddenly remembered John Dunn Hunter and my desire to learn more of his life and times. I discovered that the *Memoirs* cited in the butterfly guide was in the library and

I wrote the code letter or number on a slip and presented it at the desk in the center of the room. There I was informed that the book I was asking for was kept in a special collection behind a locked door but that I could gain admission by applying to an official down the corridor. I applied. The permission was granted. I knocked on the locked door and it was opened unto me when I displayed the pass I had received. The famous *Memoirs* came down off the shelf, and where I sought enlightenment I found only mystery. John Dunn Hunter told of his youth and early manhood with the Indians and that was practically all there was to the little volume. There was nothing about going to Europe and making a fortune, or how he became a cultured gentleman, a scientific collector and a friend of the King of England. The book was handed back to the librarian, the door was unlocked to let me out and I departed a baffled man.

But it stuck in my mind and some years later, when I was in the American Museum of Natural History, I spoke of the Hunter story to Dr. C. H. Curran, Associate Curator in the Department of Insects and Spiders. Dr. Curran said it was indeed a strange tale, all new to him, and that he would look into it. Within a few days I received a letter from him that ran as follows:

Fabricius named a butterfly *Papilio huntera* in 1775 and described it accurately without ever having seen it. He based his description on two illustrations in a publication by Drury (1773). Drury already had named the insect *virginiensis*, but Fabricius ignored this, as some of those old boys so often did. Hübner, in the 1800s, called the insect *hunteri* and was followed by practically everyone since that date. There was a famous physician and anatomist named John Hunter in England—his brother William was a noted surgeon, too—but it is quite unlikely, in view of the spelling, that Fabricius named the insect after anyone in particular.

One thing is certain—it was not named for John Dunn Hunter because Fabricius never heard of him. Fabricius died in 1808. This John Dunn Hunter turned up in New Orleans in 1818 and could not speak English. In 1823 he published his *Memoirs* and about that time (it was said) was "lionized in England." He was killed in 1827 by an Indian—in Texas, I think. He never was a scientist but was, apparently, a monumental liar and troublemaker. Now I'm wondering who wrote that butterfly book in which you read the fantastic story you mentioned.

The curious thing is that the fantastic story about John Dunn Hunter and his alleged connection with Hunter's Butterfly was written by the eminent Dr. W. J. Holland! When I looked again, I discovered that my tattered little *Butterfly Guide* that I had purchased so many years ago for $1.25 was the work of "W. J. Holland, LL.D., Director of the Carnegie Museum, etc., etc." If Dr. Holland had so much to say about John Dunn Hunter in a small book, I wanted to know how much more he had to say in his big volume on butterflies, so I picked up the hefty *Butterfly Book* of 500 pages "Illustrated by Seventy-seven Plates and Numerous Figures in the Text, Giving Over Two Thousand Representations of North American Butterflies," and on Page 154 I found a brief description of Hunter's Butterfly

208

(under a different scientific name; it's *Pyrameis huntera* in the little old book and *Vanessa virginiensis* in the newer and much larger book) and no mention of John Dunn Hunter at all! Not a single word of the entrancing story! The little book was published in 1915. My edition of the larger volume was dated 1931. Evidently in the interim somebody had told the eminent entomologist that either he had mistaken his man or he had been put upon. *Quandoque bonus dormitat Homerus.*

One day in early September when Edwin Way Teale was with us in the swamp he said in a commanding voice:

"You fellows keep your eyes peeled in all directions for Monarch Butterflies. I'm trying to get statistics on their Autumn migration. Every time anybody sees a Monarch, sing out!"

It's quite possible, of course, that Mr. Teale actually was interested in keeping tab on the number of Monarch Butterflies that were flitting through our sector that bright September day, but it is most certain that he saved himself a lot of annoyance when he set us to scanning the sky and landscape to count migrating Monarchs. It didn't allow us much chance to inspect the ground under our feet or the shrubbery around us for strange insects in the winged or larval stages that we might ask him to identify. As a matter of fact, it turned out to be a rather poor day for Monarchs. The weather was wonderful for man, beast and butterflies, but we counted only thirty-eight Monarchs on our walk. Two days earlier, over only a portion of the same trail, more than fifty Monarchs had been sighted and we hadn't been keeping as strict a watch. But Mr. Teale seemed satisfied with the count and he caught one male Monarch—without harming it—to show us the black "pockets" of scent scales on the hind wings. These pockets were supposed to give off a sweet odor. We all took a sniff. I thought I detected a faint odor of something like snuff, but perhaps I only imagined it. Mr. Teale retreated to his Long Island fastness and from there he wrote under date of September 12:

Yesterday and today the Monarchs have been going through this region in vast numbers. I took moving pictures of them clustered by the hundreds on oak and Tupelo trees. I counted 100 in the space of a foot on one branch. We noticed an interesting thing. Butterflies flying to the east of the grove—the breeze was from the west—would make almost a right-angle turn when opposite the place where the congestion of Monarchs was greatest and fly toward the trees where the other butterflies were clustered. Possibly the scent of the massed males is the thing that guides the main body of the migrants. As you know, bees in a cluster at swarming time send out a scent signal to attract bees still in the air. It is interesting to note, also, that on the flight south the males and females are said to be about evenly divided while, according to Dr. Austin Clark of the National Museum, when the Monarchs return north in the Spring, females predominate. The insects fly south in a vast straggling swarm. They come northward individually rather than in a mass. Perhaps it is the greater percentage of males —hence a stronger scent for group collection—in the southward migration that causes this difference in density in the northward and southward flights. It's just a conjecture—the Teale hypothesis for the day.

210

Despite the earnest efforts of Mr. Teale to instruct us, we learned little about *Lepidoptera* and less than that about the myriad other forms of interesting insect life around us. I remember one sorrowful experience of my own connected with insect night life in Connecticut. Of a Summer evening I was taken to a rambling house on a hill in the town of Weston on the pretext that good singing would be heard there. The singing was good, as advertised, but after the musical program was over and the guests were stuffing themselves with food and drink, I saw an uninvited company of moths clinging to the outside of a window screen, fluttering wildly in a vain attempt to beat their way through the screen into the room to reach the light by which they had been attracted. One of this uninvited company was a large moth of more than two inches wing spread and so vividly marked with pink and black on the hind wings that I thought it would be no trick at all to keep the pattern and color in mind and identify the moth by looking through my moth guide when I reached home. I said as much to the assembled multitude that gathered around me when I called attention to the fluttering group on the window screen, and I rashly promised the host of the occasion that I would send a letter giving the name of this distinguished nocturnal visitor with the gaudy hind wings.

When I reached home I took down *The Moth Book* by Holland and found, to my horror, about three pages of colored pictures of moths that much resembled, one and all, the particular moth that had attracted my attention under cover of darkness in Connecticut. They were of the genus *Catocala,* and Dr. Holland wrote of them:

"Over one hundred species are attributed to our fauna. Of these the majority are figured in our plates."

They were, indeed, and I thumbed backward and forward and pored over text and plates for more than an hour without coming to any specific conclusion. The moth remains in my memory as just another mystery of Summer night life in Connecticut.

15

I am but mad north-northwest; when the wind is southerly, I know a hawk from a handsaw.
                                        HAMLET, ACT II, Sc. 2

It was September 10 when we suddenly realized that Summer had passed and we were walking down a road in typical Autumn weather. What brought it sharply to our attention was a cry from the Medical Student, who, as he gave tongue like a hound on the trail, pointed to the clear blue sky overhead in which three Broad-winged Hawks were circling and drifting steadily to the southward. They were the first of the hawk migrants of the Autumn that we had seen, and the heralds of wonderful weeks to come—not the English Autumn, "season of mists and yellow fruitfulness" of which John Keats sang, but the superb and radiant Autumn of North America, when the woods and fields are covered with color and the weather is clear and sparkling for days on end.

We looked around us and saw the roadside decorated

with asters of many species and varying shades. Across the valley we could see on the far hillside the shining red patches that told us, even at that distance, which were the Sour Gums among all those trees. The sumacs were tipping their leaves with scarlet and the Flowering Dogwoods were turning a deep crimson. Autumn was on the premises, but who had seen it coming over the horizon?

To my eyes the first outward and visible sign of Autumn in our neighborhood is displayed by the Tulip tree, the lofty green foliage of which becomes spotted with gold in late July. It's a striking note amid the Summer greenery and is bound to catch a roving eye. The first Autumn sounds to reach my ears are the lisping notes of traveling warblers among the trees and the "chink-chink" of southbound Bobolinks passing high overhead—invisible but audible. Some of the shore birds begin to move southward in July and, on higher ground, the Orchard Oriole doesn't pause to fold its tent like an Arab but as silently steals away. Redstarts become restless about the last week in July. We hear them *tsip-tsipping* all around us on our walks and we catch glimpses of them—mostly females and young of the year —as they flash about the shrubbery and the lower branches of the forest trees. I saw the Louisiana Water-thrush on our ridge in late July and, since it does not breed in the vicinity, it was evident that it was a bird getting an early start on its Autumn pilgrimage.

In August the Joe-Pye Weed comes into color and the Purple Grackles flock noisily in the groves. The Gray

Squirrels rob the Hazel-nut bushes while the nut clusters are still green and the male Scarlet Tanager is seen wearing a mottled coat in which green is displacing the bright scarlet of nuptial glory. There is a wide movement of warblers in the woods and through the roadside trees and we note the passing species: the Redstart in astonishing numbers, the Canada Warbler, the Black and White, the Worm-eating, the Magnolia, the Northern Yellow-throat that lurks at a low level in the shrubbery and the Parula Warbler that gives us a pain in the neck because we have to twist our heads back so far to find it in the upper reaches of the trees. In late August and early September the roadsides and meadows are gay with goldenrod and asters of many species and, in the woods, we find the smaller flycatchers—the *Empidonax* group—traveling in company with waves of warblers, but not always in perfect harmony. I have seen some of the little flycatchers striking at warblers and some of the warblers swooping at perched flycatchers, but whether it was done in fun or earnest I could not tell. At any rate, I never saw any damage sustained on either side in such assaults.

Insect life seems to reach an overflowing peak at this season of the year, and even sparrows, which are seedeaters by preference and tradition, may take to a diet of insects when it is so abundantly offered. We have seen Chipping Sparrows darting up into the air and grabbing flying insects in their bills as neatly as any flycatcher could perform the feat. Along the river road in early September I saw migrating warblers striking at the foliage of

215

trees in such a way that whitish flying insects resembling small moths were driven out almost explosively and went fluttering off with the warblers in zigzag pursuit. The insects were all over the place and some of them, exhausted or injured, fell on the roadway. I picked some up and found them to be pale green in color with a covering of whitish powder that made them look much lighter on the wing. They were poor flyers, whirring along more like grasshoppers than artistic air pilots, such as moths and butterflies, and they looked rather tough of texture to be food for warblers. But perhaps tourists have to be content with anything they can get in the way of food. Since the pale green insects were utter strangers to me, I packed off a few to the American Museum of Natural History where Dr. Willis J. Gertsch, Associate Curator of the Department of Insects and Spiders, identified them as *Ormenis septentrionalis,* one of the Fulgorids or Lantern Flies.

Now the Red Maples in the swamps are turning a brilliant color and the White Ash leaves are beginning to fade to a dull mahogany. The Wild Bean and the Hog Peanut are in bloom along the stone fences, the seed pods of the Spotted Touch-me-not are ready to pop, the Savory-leaved Asters appear as a light blue area in the rich brown expanse of an open hillside—and three Broad-winged Hawks on southward migration overhead tell us that Autumn is not only upon us but all around us, that Summer is now a memory. There is a tang to the air that lends a lift to the traveling foot. There is a brightness to the landscape that puts elation in our hearts. It is good to be walking in such surroundings. It must have been just such September scenery that roused Bliss Carman to write:

> *Now the joys of the road are chiefly these:*
> *A crimson touch on the hardwood trees;*
> *A vagrant's morning, wide and blue,*
> *In early Fall when the wind walks, too;*
> *A shadowy highway, cool and brown,*
> *Alluring up and enticing down*
> *From rippled water to dappled swamp,*
> *The outward eye, the quiet will,*
> *From purple glory to scarlet pomp,*
> *And the striding heart from hill to hill . . .*

Once we find that the hawks are on southbound flight, we take to the higher ground so that we may have a good look at them as they pass along in their flight lanes. On September 17 we went by car to our "Lake District" and then walked along a wood road on a ridge east of a large lake. The wood road ran through a slash or cutover section of what had been a good stand of oak, and the scrub coming up gave thick cover for the wildlife of that area. We saw many deer tracks in the dust and mud of the old road, and a native of the regions whom we met on our march told us that there were more foxes than sour grapes around there this year. He said it as though his teeth were set on edge thereby. He admitted that he was dead set against foxes. He said they killed Pheasants, a dastardly deed when done by foxes. He had planned to kill those Pheasants himself, but the foxes were a little ahead of him in the field.

We hoped to see some southbound hawks, but the wind was not blowing in the right direction and we saw only a few, fourteen in all, mostly *Accipiters*. One sandy section of the old wood road we followed was swarming with our common native Crickets, though why they were so plentiful in this spot was a mystery to us. Further along we saw a Wood Frog hop off into the underbrush and we took after it to catch it. Wood Frogs are lively jumpers and this one escaped our clutches in the confusion of the chase, but we started another one in the woods, caught it and turned it loose unharmed after we had admired its trim lines and the neat black patches on the sides of its head.

There were plenty of dull-colored Robins, gaudy Flickers and clamoring Blue Jays in migrating groups working their way over and through the slash, but most of the warblers already had passed through and we heard only a few chips of late travelers. We noted a Mourning-cloak Butterfly in prime condition and thought perhaps that it would spend the Winter in some

hiding place in the area. It was astonishing to think that such a delicate creature as that butterfly could live through the snow and rain and icy winds of our Winters, but some Mourning-cloaks do it each Winter and, after such rugged hibernation, we find them on the wing in the leafless woods of early Spring when March more or less makes sweet the weather

> *With daffodil and starling*
> *And hours of fruitful breath.*

We had a small boy with us on this trip and he called out to us that he had found a "strange bug" on the road. He was down on his hands and knees looking at the "bug" and he asked us what it was. Any insect expert could have told him in a jiffy, but we had no insect expert with us, so I studied it under my little magnifying glass with the idea of looking it up when I reached home. It was a beetle of some sort about an inch long with a striking pattern on its back, odd enough to raise the hope that it wouldn't be much trouble to identify if it was illustrated on any plate I had in reference books in my library. The hope was justified when I consulted my Lutz *Field Book of Insects* in armchair comfort later in my ordinary unscientific fashion. I glanced over the colored plates in the back of the book and there was our insect to the life under the name of *Cicindela generosa,* which is listed by Lutz as one of the common Tiger Beetles of our region. It may be common to insect experts, but it was the first time that I had seen eye to eye with it and, through a magnifying glass, it was truly an astonishing creature. If these Tiger Beetles came as large as Newfoundland dogs they would scare the wits out of everybody in this section of the country.

The Artist and I headed for the Berkshires the next day in the hope of finding more hawks on the move. We knew that the top of Mount Tom was a famous observation point during the Autumn hawk migration, and Homer Newell of Huntington, Mass., whose hobby is the taking of motion pictures in color of wildlife, drove us up the winding road along the mountainside at 9 A.M. on September 19. The mist was still in the valley and even part way up the slope of the mountain at that time of morning, but a warm sun soon cleared the landscape and we had a wonderful view as we perched ourselves on a rock ledge at the northerly end of the Mount Tom ridge. The "ox-bow" of the Connecticut River was just below us, and beyond it lay Northampton, one of the attractive college towns of New England.

We saw hawks migrating on either side of our ledge, but the disconcerting thing was the way some birds seemed to pop up almost under our feet and whip past us before we could be sure what species they were. We thought that would be a great place to put up a stuffed Great Horned Owl on a pole so that the hawks would make it a stopover point to heckle the owl. I report with regret that, on a later occasion, we carried out this plan of putting a stuffed Great Horned Owl on a pole on this part of the ridge and no hawks came anywhere near us or the defunct owl. On this September day, however, we had Broad-winged Hawks, Sparrow Hawks, Duck

Hawks, Ospreys, Sharp-shinned Hawks and Cooper's Hawks within range of our field glasses. Most of the migrants—by far the greater number—were Broad-winged Hawks. We saw a group of observers on a steel tower platform not far away and, when we joined them, we found that their count was already in the hundreds. There were intervals when no birds were in sight and there were moments when fifteen or twenty circling Broad-winged Hawks could be seen at the same time. I heard later that the group on the lofty observation platform "logged" more than five hundred hawks that day, but I preferred our rocky ledge to the northward because the hawks came in closer at that point. Not only that, but with one man on watch on the ledge the others could stroll around and search for strange flowers or shrubs until a hawk was sighted. It was in this fashion that I found the Pale Corydalis blooming in rock crevices near the ledge that day. It was a new flower for me and, even without hawks, it made the trip notable. And the day that we put up the stuffed owl in vain and no hawks came near us on the ledge, the vicinity was alive with Red-breasted Nuthatches. There is always a profit balance at the end of a day in the open.

On September 20 we were back in our "Lake District," perched on a ridge and looking northward over a sweep of rolling ground. It was a fine sunlit morning, but we could have done with a bit more wind out of the northwest to help the hawks along. Favoring winds and rising currents of air (thermals) help the hawks no little on their long migration flights, and when conditions are just right the birds really pour southward along certain flight routes. We were out early this time and Herman the Magician had a hamper of tomatoes from his own garden with him. The Medical Student had a bag of McIntosh apples and a thermos bottle filled with hot tea. We were going to make a day of it. The side of our observation ridge was thickly covered with White Pine, but the summit was an open field and there Herman rejoiced exceedingly when he found Nodding Ladies'-tresses in bloom. He pointed to them in triumph, but I'm afraid I was not enthusiastic. They were not much to look at in my opinion.

"Sir!" thundered Herman. "They are orchids!"

Perhaps so, but I much preferred something more pleasing to the eye even if of more humble stock. I loved the Common Dandelion, for instance, and the Common Daisy. If the Common Dandelion were rare and expensive, how it would be praised for its matchless beauty! But it grows everywhere in our region and I have seen it bravely blooming eleven out of twelve months in the year, wherefor—as Prince Hal said of the wisdom that cries out in the streets—no man regards it. But when I looked down on Herman's mediocre representative of an aristocratic family, he couldn't have been more indignant if he had been charging me with treason. Though ordinarily one of the most generous of souls, at that moment I think he would have refused me even one small tomato out of the full hamper of blushing beauties that he had brought with him to stay our hunger while we were on the watch for hawks. But Herman is no man to nurse a grudge for long and, hawks being one of his four furious pursuits, he was soon sweeping the sky to the northward with his field glasses and, indeed, the first one of the group to sight game.

"Here comes one now!" he said as he peered through his glasses. "See that barn on the ridge? Go straight up from there, about thirty degrees."

We looked and found the hawk. It was only a speck in the sky and never did come near us. It passed far to the eastward, quite high.

"Might have been a Sharp-shinned," said Herman.

"Could have been a Cooper's," said the Medical Student.

I said it might have been either, but we never would know. Then Herman let out a whoop.

"A big one—I see white on its head—I think it's a Bald Eagle!" he cried.

It was, indeed, a large bird that was flapping slowly toward us and certainly there was a glint of white up front, but when it came closer we saw that it was an Osprey. Hardly had it passed when the Medical Student sang out:

"I see one—two—wait a minute—five altogether—coming over that stone house on the ridge. They're circling and soaring—Buteos for sure—can't tell the species yet——"

On they came, circling lazily but somehow drifting swiftly southward high overhead at the same time. They must have been a thousand feet above us when they passed directly overhead and we saw that they were

223

Broad-winged Hawks, as we had expected. We counted on seeing far more of them than any other species on these September flights. However, we did see a few Cooper's Hawks and Sharp-shinned Hawks as well as half a dozen Sparrow Hawks, one Red-shouldered Hawk and a mature Bald Eagle that came quite close to us. This bird was in great feather. It looked as though some professional bird trainer had given it a wash and body polish for the trip. Its yellow bill was glistening with color and its head and tail feathers were shining white in the clear blue sky.

There was a lull for a time after the Bald Eagle went by majestically, and we ate the apples and tomatoes, smoked, talked and kept an eye to windward at the same time. The hawks do most of their flying in the morning. The traffic slows up around noon and then picks up again, though the afternoon flight doesn't often measure up to the morning numbers. Herman became drowsy and the Medical Student—whether to keep him awake or put him to sleep I could not be sure—sang:

> *"Good King Wenceslas looked out*
> *On the feast of Stephen . . ."*

No hawk came within two miles of us while the Medical Student was singing. This may have been some form of higher criticism, though the Medical Student would by no means agree with this when I suggested it. Our afternoon record was spotty—just a hawk here and there —but one was a Duck Hawk, a bird that I never see without a thrill. This is the American cousin of the

224

Peregrine Falcon of medieval history and romance. It is the avian ace of the air, the fastest thing on wings in our part of the country, a marauder, a fierce and fearless predator for its weight and size. We have seen Duck Hawks swooping at Bald Eagles, apparently just for the fun of it, and Dick Herbert, who keeps a census of all the Duck Hawks that breed along the Palisades, told us that he had seen a Duck Hawk knock down a Great Horned Owl that happened to pass too close to the Duck Hawk's nest.

There is no mistaking the Duck Hawk when it stoops to conquer. It comes down with a lightning plunge on half-shut wings. How fast it drops is a matter of estimate, and some estimates I have seen I believe are much too enthusiastic. The Falconer, when he was a fighter pilot in France in World War I, found himself flying parallel to a Peregrine Falcon as he was returning from a foray over enemy territory in the Champagne sector. He turned to test the speed of the bird and followed it down when it dived to get away from the plane. He knew the speed of which his plane—a Spad—was capable and of this experience he wrote:

"I would say that the Duck Hawk has a cruising speed of about 75 miles per hour and a diving speed of about 150 miles per hour. At least, this one did."

The Dramatic Critic told me that once on a Winter day he was watching a line of ducks winging swiftly northward just a few feet above the water of the Hudson River. For some reason the ducks seemed to be in a tearing hurry as he watched them through his field glasses. Suddenly a Duck Hawk, coming from the rear, shot past the line of ducks so fast that, as the Dramatic Critic solemnly stated, it made the ducks looks as though they were walking on the water. Once the Medical Student and I went across the Hudson River and along the foot of the Palisades in the hope of seeing a Duck Hawk. It was in early Spring and the ground was muddy, so that walking was a heavy process. We plodded along for hours and saw nothing except a few stray ducks and hundreds of Herring Gulls on the river. By late afternoon we were worn and weary and a cold wind had come up to add to our discomfort. Just as we were going down the slope to take the ferry home in the dusk, the Medical Student grabbed my arm and said: "Look!" I looked and all I saw was a darkish speck shooting downward across the sunset sky with almost incredible speed. There was just one bird in our region that could fly in such fashion—the Duck Hawk! We were dog-tired and our shoes were a mass of clotted mud, but there was a glow of satisfaction in our hearts as we went aboard the ferry. We had seen "Old Sickle Wings," which is our nickname for any Duck Hawk because of its shape in hunting flight, and we were well content.

Sometimes we didn't have to go far afield to see hawks on Autumn migration. One Saturday afternoon in late October the phone rang and it was Herman the Magician at the other end of the wire. He was excited but he held himself well in hand as he said that there was a big hawk flight in progress—they were sailing over his house in a southwesterly direction—it was a grand sight and I was urged to stir my stumps and rush outside the house to watch it. We live not far apart on the same suburban ridge and I could have seen from my lawn any hawks that were passing over Herman's roof, but I had some writing to do before going off to watch a football game at Baker Field, so I thanked him and went back to my typewriter. In about twenty minutes the phone rang again and this time it was Herman with a wild call and a clear call that could not be denied.

"I've counted forty-three hawks since I last phoned you!" shouted Herman over the wire. "Don't miss this! Jump outside and have a look or forever hold your peace!"

That did seem like a lot of hawks coming over our little ridge in twenty minutes, and work could be put aside for a spectacle of such magnitude. I grabbed my field glasses and was outside the door just in time to spot two Red-shouldered Hawks circling high over the house and drifting to the southwest at the same time. Just as they disappeared, three more came into sight from the northeast and sailed overhead. I dashed back into the house to phone the Artist, who also inhabits our ridge and could leap outside at a moment's notice. I gave the news to the Artist and then resumed my own scanning of the sky, with the result that I counted thirty-four hawks passing over the house in thirty minutes, and all that were close enough to be identified definitely were Red-Shouldered Hawks. The Artist, who watched longer, counted seventy-two flying over his territory and Herman counted more than a hundred

227

before he had to stop and go about more pressing and less attractive work.

But we are not exclusively occupied with migrating hawks in September or the bright October days that follow. We have our debates over some of the puzzling warblers in Autumn plumage and, when the warblers have gone through, we look over the migrating sparrows that are everywhere along the roadside and in the underbrush. We welcome back to our dooryards the birds that spend the Summer in the woods, the Black-capped Chickadee and the White-breasted Nuthatch that are regular diners at our baskets of suet through the colder weather. Toward the end of September the White-throated Sparrow and the Slate-colored Junco come down from their more northerly breeding grounds to spend the Winter with us. My earliest Red-breasted Nuthatch was on September 17, but this species is unreliable; one year it will come early, another year late; some years it is plentiful and other years a single individual is almost a prize in our neighborhood. On the other hand, the Ruby-crowned Kinglet and the Golden-crowned Kinglet are not only lovely little birds, but you can trust them to appear on Autumn migration about October 1, give or take a few days, and probably there will be at least one Brown Creeper with each kinglet group. The Ruby-crowned Kinglet is mostly a transient in our territory, but some of the Golden-crowned Kinglets and Brown Creepers remain as Winter residents. We have had Pine Siskins drop in on us as early as October 15 and they usually stay all Winter when they come, but they are as unreliable as the Red-breasted Nuthatches: some years they are abundant and other years we see only a scattered few in a whole Winter.

On a sunny day in mid-October we found the Fringed Gentian blooming in a wet meadow, and the fence rows were brilliant with the red berries of the Black Alder. The thickets were alive with migrating sparrows and the last of the warblers were going through, the Yellow Palm Warbler, the Black-throated Green Warbler and the Myrtle Warbler. This last species we have as a Winter resident where there are great patches of Bayberry. Just inside the fringe of the October woods we saw Hermit Thrushes moving along with quiet dignity. From the brown grass of an open hillside we disturbed several Meadowlarks that flew off with a fine display of white outer tail feathers. The landscape was red and gold and soft russet, the day was bright and warm, and down from the blue sky above us came the plaintive warbling notes of Bluebirds, always a lovely sound to our ears. October is a wonderful month with us.

16

All places that the eye of heaven visits
Are to a wise man ports and happy havens.

RICHARD II, ACT I, SC. 3

It's a curious thing that so many persons mark Indian
Summer far too early on their calendars. They place
it in late September or any part of October, but the real
Indian Summer comes late in November and may even
lap over into the first week of December. It is, in a
measure, the calm before the storm, the last fairly warm
period before the cold blasts of Winter come whirling
down over a bleak landscape as the daylight shortens
to the solstice minimum. At such times there comes an
atmospheric haze over the countryside that the old
settlers laid to brush fires set by the Indians, and thus
the belated mild spell received its name.

The leaves have gone from the trees now and we can
look clear through the woods that were solid walls of
greenery through the Summer. We see the framework
of the forests, the stout trunks, the strong limbs, the net-

230

work of interlaced branches etched against the sky, a sight on which our friend the Poet mused as follows:

In songless air
In the straight sun,
Stricken and bare,
Emerges one
Whose line of bough,
Being stripped of mask,
Arrests us now,
Who stand to ask:
Is this the same
That once we found,
Whose drowsy name
Was a Summer sound?
We need a strong
Lean name for this,
Bare of bird-song,
Stilled of leaf-kiss,
Imprinting late,
On mortal seeing,
The ultimate
Hard shape of being.

It was this same period that stirred William Cullen Bryant to write that the melancholy days had come, the saddest of the year, a sentiment with which our strolling company would by no means agree. From my younger days I remember the lines of another poet, apparently choked by something more than seasonal emotion, who tackled the topic in this fashion:

Chilly Dovebber with its boadigg blast
  Dow cubs add strips the beddow add the lawd;
Eved October's suddy days are past—
  Add Subber's gawd!

231

But we like the late Autumn just the same. The Artist sees much that is beautiful about the outdoors in all seasons. We probably see less, but still it is enough to gladden our hearts and lure us along the river road or the woodland path twelve months of the year. We wear old clothes and are never daunted by bad weather. Not snow, nor rain, nor heat, nor gloom of night stays us from the leisurely completion of our unappointed rounds.

For that matter, and since we are down to November, this month has some fine points in our region. The Artist remarked—as we rambled over the rim of October into the foreground of November—that we were just coming into the Blakelock period by hill and dale. Those who meander through the art museums of this country will recognize the artistic touch in that observation. It struck a responsive chord in me because Ralph Blakelock is one of my favorite American painters. But the finest Autumn landscape from the hand of Blakelock— or Rembrandt or Hobbema or Théodore Etienne Rousseau—lacks the depths of color, the sweep of vision, the evanescent beauty against a background of eternity that Nature paints so lavishly for us on a grand scale and exhibits free of charge for weeks at a time.

I set about listing the wild flowers that we found braving the low sun and the chill winds of November in our territory. One of our prizes was the Fringed Gentian—

Thou blossom bright with Autumn dew
And colored with the heaven's own blue—

and on the third day of this eleventh month I found two violets—I took them to be *Viola papilionacea,* though I may have been wrong about the species—modestly spreading their petals in the fading grass at the side of a footpath that we often tread. There were, of course, asters in bloom in the woods and fields, but they run to species too numerous to mention and too hard to tell apart so far as I am concerned. I can distinguish about half a dozen of the more common species and the others I leave to the Astronomer, who goes over them with his magnifying glass and ultimately pins them down one way or another. Our goldenrods are in the same category—many species hard for me to distinguish, but the handsomest to my eye is the Rock Goldenrod. Of the aster group I think I would give first prize for beauty to the New England Aster.

We found November fields dotted with shining yellow Dandelions and we came upon meadows that still offered us Red Clover in full rich flower. I must register distinct approval of Red Clover. It is beautiful; it is fragrant; it is good for the soil; cattle relish it; bumblebees feed from its florets; Rabbits munch it with zest; it gives a fine touch of color to the meadows and it flourishes merrily from May to November. It is one of the common things—like air and water and sunsets and the march of the constellations at night—that we haven't sense enough to appreciate no matter how essential or beneficial or decorative they may be.

The Evening-Primrose was easily added to our November list and we also found some blooming stalks of Butter-and-Eggs and a few pale blue flowers of the roadside Chicory. On the fringe of an old road through the woods I found White Snake-root in sturdy bloom. Apparently this is the hardiest of the genus *Eupatorium* in our area; it holds up its flat-topped white clusters long after Joe-Pye Weed and Boneset have faded out and withered away. Other items for our November flower list were a few sickly yellowish blooms of the tall Wild Lettuce, an assortment of Beggar-ticks, and some Lamb's Quarters that I detected in flower with the aid of my little magnifying glass, which perhaps was cheating a little.

Those were my own contributions to our November bouquet of wild flowers. I called on other members of our road company for help and it was the Artist who helped most. The man has an uncanny eye; nothing on the ground, in the thickets, on the branches of trees or in the sky overhead escapes him. In no time at all he found Witch Hazel in bloom in a moist patch of woods, and along the roadside he found Bouncing Bet, Wild Carrot, Yarrow and a lovely Common Daisy, as fine and healthy a specimen of *Chrysanthemum leucanthemum* as June ever offered for observation in our territory. But his prize of that November day was one that he brought up from a ditch—a Wild Geranium (that Messrs. Britton and Brown prefer to call Wild or Spotted Crane's-bill) in the purple-pink of perfection.

It was just a few minutes after this great discovery that the Artist began to tell us of an odd animal he had seen the past Summer in the Berkshires. He said it was a Fisher—or Pennant's Marten, as some name it—an animal of the *Mustelidae* or Weasel Family that is quite valuable for its fur. The Fisher is larger than a Mink and smaller than an Otter. It might pass, in a hurry or at first sight, for a short-legged, dark-colored fox. It

was the low-slung body of the animal that attracted the attention of the Artist. The Fisher, by the way, is not appropriately named. It does not make a habit of going fishing, although it will eat fish or—like Mark Twain's famous Stolen White Elephant—almost anything else it comes upon, including Squirrels, Rabbits, birds, eggs, and even Porcupines, which does not argue a discriminating taste. It climbs readily and breeds in hollows in trees well above the ground. Constant trapping has made Fishers scarce over their natural range, and the little animal works mostly on the night shift, anyway, so the report of the sight of a Fisher by daylight in the Berkshires bordered on the sensational. In fact, it was too much for Herman the Magician, our Big Game Hunter, to swallow easily. He halted abruptly when the Artist had told his tale and asked him whether or not he was also seeing snakes when he saw the Fisher. The Artist is a notoriously sober person and he indignantly rejected the base insinuation that his vision had been blurred by alcohol in any form or quantity.

"Well, there hasn't been any other report of a Fisher in that region for the last fifty years," said Herman coldly.

"Stick to your orchids and your owls, my fine-feathered friend!" said the Artist. "I'll give the lectures on wild animals."

He went ahead with his talk and told how he had seen the Fisher twice. The first time he had just a quick glance at it in the late afternoon when it crossed a road ahead of him. He knew all the wild animals that were regular inhabitants of his area of the Berkshires and this was distinctly an odd one to him. He suspected that it was a Fisher, largely because he thought it couldn't be anything else. The next time he saw it was in the twilight when he was driving along a main highway. The odd animal was going down a rock-studded embankment on the east side of the road and the Artist, who had passed it at thirty miles per hour, stopped the car up the road and walked back to investigate. From the road he looked down over an array of huge rocks that bolstered the shoulder of the road along the side-hill at that point, and on one of those rocks, about forty feet from the Artist, was the strange animal. The Artist had a fine view of it for fifteen minutes as it prowled among the rocks, looking for mice or Rabbits or anything else to eat. It was a Fisher without doubt, according to the Artist.

"Let's ask the man at the museum about this," said Herman stubbornly.

A few days later Herman and I were sitting around a table in the American Museum of Natural History with some of the noted scientists of the staff and I presented the case for the Artist, who was unable to be with us. Herman offered his objections to the story. Dr. Harold E. Anthony, Curator of Mammals and editor of *Mammals of North America*, listened to us and then gave his opinion. He said that Fishers ordinarily were found somewhat north of the spot in the Berkshires where the Artist had reported the animal, but originally they were native to all that region and certainly it was not illegal for a man to see a Fisher where the Artist had reported one. Was the Artist a competent outdoor observer, a steady man in the field?

"None better!" admitted Herman heartily.

"Then I suppose he saw a Fisher," said Dr. Anthony calmly.

I know he did. If the Artist said he saw something, I'll wager that he did and I don't care how far off it was, or whether it was night or day, hot or cold, in sunshine or rain. He has the keenest pair of eyes that I ever encountered.

One mid-November day we went to Bear Mountain to see what the Beaver in the park there were doing. As an added starter we had our friend Dr. C. H. Curran, the entomologist from the American Museum of Natural History who had given expert advice to me in the matter of Hunter's Butterfly. The first thing we saw outside the little museum in the park was a Golden Eagle tethered to a log. While we were looking at the big bird, a Bald Eagle flew across the sky at a far height and the captive Golden Eagle, spotting it immediately, turned its head to follow the flight of the free bird until it disappeared to the westward.

In the little zoo at the park there were two dripping Beaver that had been caught designedly in box traps. There were so many Beaver in the Bear Mountain area that the park officials were trapping some to be given away to other parks and zoos. One of the Beaver we saw was a big fellow that weighed sixty-eight pounds, but the top record for one trapped in the park was seventy-two pounds. We went up into the hills to inspect a Beaver dam and adjacent lumbering operations of the animals. We came upon one large dam with half a dozen small dams on the stream below it and we saw, in the large pond they had created, their houses with an anchored supply of juicy food sticks close at hand.

The amiable legend that these animals fell a tree so

expertly that it always falls exactly where it best suits their purposes is still current, but when we looked over the little hillside that had been lumbered for this particular dam, we saw that the trees had fallen every which way and many that the Beaver had been unable to drag off were lying on the ground in disorderly directions. Dr. William H. Carr, former Director of the Trailside Museum in Bear Mountain Park, who put in some time studying these animals, once wrote me that "trees cut by Beaver fall in every direction, sometimes on the Beaver"! Scientific reports have ruined the reputation of the Beaver as an engineer and I rather question the expression "working like a Beaver." The animal loafs a good part of the Summer and apparently does nothing all Winter except eat and attend to a few minor repair jobs around the house. All in all, I think that *Castor canadensis* is an amiable and interesting animal but an old fraud as an expert engineer or pious example for lazy human beings.

We took note that the Beavers had no particular taste in trees. They gnawed down oak, poplar, birch, ash, maple and other kinds indiscriminately. We lingered around the pond they had made but saw no animals "in person," though we went so far as to stand on a housetop and knock to attract attention. Not as much as one small Beaver poked out a nose that we could see, but in a scramble around the dam we came close to a small brown bird that was flitting about in the matting of logs and brush, and when we looked through our glasses we saw that it was a Winter Wren, which is always a welcome sight. As we walked down a park road a young White-tailed Deer dashed out of a thicket, across the road and up a steep hillside. It turned on the crest of the ridge to survey us, and off to the right we saw a magnificent buck with a fine spread of antlers going over the skyline.

We crossed an open field and went into a patch of woods that contained some old buildings, and it was there that the entomologist went to work. He took us inside one of the rickety sheds and showed us mosquitoes by the dozen on this bleak November day—two different species of *Anopheles,* he said—clinging motionless to the woodwork in the darker parts of the empty building, all set to pass the Winter in that fashion. Who would think that so frail an insect as a mosquito could live through a Bear Mountain Winter in an empty house with the doors open to the freezing wind and the low temperature of that area! The Black Bear of the same region, protected by a thick fur coat, holes up in a hollow tree to survive the bitter weather, but the fragile mosquito has no fur coat as a protection when it spends the Winter behind a door in an empty house on Bear Mountain. This contrast is just one of the myriad mysteries in Nature.

Our visit to the Beaver community reminded me that it was time to resume the cutting of wood a little closer to home. The Artist and I pair up to furnish wood for open fires at our houses as soon as cold weather comes on. We have a two-handed saw, an ax, a sledge hammer and four steel wedges. We began some years ago with

240

a Red Oak that a hurricane had blown down near the Artist's house. It was about four feet in diameter at the butt and we were all Winter sawing and splitting it into fireplace size. The following year we branched out and cut up trees that had fallen on the land of neighbors. We gave the owner of the property half the wood and the Artist and I each had a quarter of the end product for our efforts. It was nice work in clear cold weather and we enjoyed ourselves with the different species of trees that we dissected—with one exception! That was an American Elm—or a huge branch from a tree of that species—and it was by far the toughest thing we tackled with our wedges.

The Artist and I share a deep enthusiasm for the American Elms of the New England area and I love that part of *The Autocrat of the Breakfast Table* in which Oliver Wendell Holmes, who charted all the great trees of New England and visited them regularly to see how they were getting on, told of his hunt for a noble American Elm. Wait a minute—judge for yourself:

As I rode along the pleasant way, watching eagerly for the object of my journey, the rounded tops of elms rose from time to time at the roadside. Wherever one looked taller and fuller than the rest, I asked myself—"Is this it?" But as I drew nearer, they grew smaller—or it proved, perhaps, that two standing in line had looked like one, and so deceived me. At last, all at once, when I was not thinking of it—I declare to you that it makes my flesh creep to think of it now—all at once I saw a great green cloud swelling on the horizon, so vast, so symmetrical, of such Olympian majesty and imperial supremacy among the lesser forest-growths, that my heart stopped short, then jumped as a hunter springs at a five-barred gate, and I felt all through, without need of uttering words—"This is it!"

I call that wonderful because the old Autocrat so well expressed what all of us who love fine trees feel about them. But there is another side of the *Ulmus americana* that I regret to expose to view. The Autocrat, indeed, more than hinted at it when, in constructing *The Deacon's Masterpiece, or The Wonderful One-Hoss Shay,* he described part of the material as follows:

> *The hubs of logs from the "Settler's ellum",—*
> *Last of its timber,—they couldn't sell 'em,*
> *Never an axe had seen their chips,*
> *And the wedges flew from between their lips,*
> *Their blunt ends frizzled like celery-tips.*

That's more like it when a man has his feet on the ground and is trying to split a segment of American Elm. The Artist and I had no trouble sawing the branch, which was a mere fifteen inches in diameter and nothing at all to some of the Red Oak and Black Oak trunks through which we had pulled our saw. We made short work of reducing the American Elm branch to hearth-size cylinders and there remained only the job of splitting them. From an end view, these cylinders looked as though they were made of two totally different kinds of wood. The inner cylinder—the greater part of the log—was quite dark; the outer ring, a three inch belt from the bark inward all around, was reasonably light in color and looked all the lighter by contrast with the dark center section. We brought out our sixteen-pound sledge and four battered steel wedges and the Artist struck the first blow.

He drove the wedge down until its head was flush with the top of the cylinder of wood. For splitting purposes he might just as well have driven a nail in the same place. There was no sign—not even a faint hint— of the block splitting. The Artist can be a stubborn man upon sufficient provocation and he prides himself on his ability to work with wood. As a youngster he was apprenticed to a master in the art of wood-carving and he still does wood-carving to amuse himself in his leisure hours. He took another wedge, tapped it in about halfway between the center and rim of the cylinder and drove it flush with mighty whacks. Nothing happened except that the first wedge, jarred by the thumping of the second wedge into the block, slowly came up out of the cylinder and was recovered by the Artist. But to recover the second wedge we had to take an ax and hew it out.

We had, over four or five Autumn and Winter seasons in succession, more or less cheerfully and at times joyfully cut and split oak, ash, maple, hickory, mulberry and assorted timber of other kinds. We had taken the gnarled and knotted sections of oak trunks and reduced them to proper size for open fires. We refused to be baffled by little cylinders of American Elm. When we couldn't split them by standing them up in regular and decent fashion, we put them on their sides, inserted the wedge midway along the length of the cylinders and fairly wrenched the wood asunder in that fashion. Except for that incident, our sawing and splitting operations helped to keep us in good health and high spirits through the cold seasons and, like Thoreau's wood chopper, we enjoyed the company of Black-capped Chickadees and other Winter birds that came to watch us at our work.

On the last day of November I was alone in our Lake District late in the afternoon and, as I walked through a grove of White Pines to the edge of one of the inlets, I frightened a flock of ducks that had been resting on the water. When they swept past me and lifted themselves in swift flight against the sunset sky, I saw that they were American Mergansers and, with their long necks outstretched, the picture reminded me of the lines of John Masefield:

> *Twilight. Red in the west.*
> *Dimness. A glow in the wood.*
> *The teams plod home to rest.*
> *The wild duck comes to glean.*
>
> *O souls not understood,*
> *What a wild cry in the pool;*
> *What things have the farm ducks seen*
> *That they cry so—huddle and cry?*
>
> *Only the soul that goes.*
> *Eager. Eager. Flying.*
> *Over the globe of the moon,*
> *Over the wood that glows.*
> *Wings linked. Necks a-strain,*
> *A rush and a wild crying.*

That was the picture I saw against the cold twilight of the last day of November. John Masefield, as a young man, had worked in a carpet mill not many miles away. He had walked that countryside on occasion. Perhaps some similar sight in the November twilight of a vanished year had stirred him to write the lines that I had been stirred to remember.

17

At Christmas I no more desire a rose
Than wish a snow in May's new-fangled shows;
But like of each thing that in season grows.
Love's Labour's Lost, Act I, Sc. 1

It's in December that the birds begin to patronize our
dooryard feeding stations extensively. From early April
until late November they have little trouble foraging
for themselves in our territory, though they do appreciate a birdbath on a lawn and will make good use of it
daily. But food is no lure for birds of our neighborhood
until December rolls around each year. Then the Black-
capped Chickadees, the White-breasted Nuthatches
and the Downy Woodpeckers come to the suet baskets
and the finches and sparrows come to the feeding trays
on which we spread a grain mixture for them.

The Artist had quite a time with the first feeding tray
that he set up on a balcony overhanging the brook that
runs beside his house. It was a nice spot for a feeding

station—running water below, shrubbery for cover nearby and an elevated tray that protected the birds from stray cats or dogs that might be wandering around. The Artist expected distinguished visitors right away at such a choice table and he sprinkled it with the finest assortment of seeds that he could buy. Then he sat just inside a window working on some sketches and waiting to see what first appeared as his guest on the balcony. To his horror, it was a House Sparrow that suddenly popped up from the shrubbery, landed on the edge of the tray and began to feed with great satisfaction.

It was not for House Sparrows that the Artist had braced the feeding tray on the balcony railing and covered it with choice seeds. He rushed out on the balcony and shook his fist at the bird as it flew away. A few hours later he looked out the window and saw not only that the House Sparrow was back but it had brought some boon companions to the feast, all of the same vulgar species. Impudence! The indignant Artist stepped out on the balcony in a hurry and gave them what is technically known in impolite circles as "the bum's rush," but he couldn't keep chasing House Sparrows all day; he had a canvas on the easel—a landscape—and he was wrestling mightily over the color and composition as he did with all his pictures in the making. When he became absorbed with his brush, the House Sparrows hopped joyously about the feeding tray and made free with the seeds. There was nothing that the Artist could do to prevent it except take down the feeding tray or give up Art, and he was reluctant to go to either extreme.

Then one day a Song Sparrow showed up among the vulgar House Sparrows. A few days later three White-throated Sparrows joined the group at the table. The Artist then began to appreciate the worth of those feathered guttersnipes, the House Sparrows; they had become decoys for his table d'hôte. Other and better birds in the vicinity heard them chuckling and chattering over the free lunch and came cautiously to see what was on the table. The strangers lingered timidly in nearby trees and shrubbery until they saw that no harm befell the gobbling and gabbling House Sparrows on the feeding tray—no guns were fired—no cats leaped out at them with terrifying claws—so the Song Sparrow and the White-throated Sparrows moved in a few feet at a time to see what the House Sparrows were tucking away with such gusto.

The next distinguished visitor was a Red-eyed Towhee, which is a common enough bird in the shrubbery around the Artist's house from early April to late October, but this was in December and, as a matter of fact, the bird stayed around the feeding station all Winter. This was decidedly out of the ordinary and the Artist was quite puffed up about it. The Towhee remained the prize exhibit—visitors at the Artist's house were encouraged to look through the window at the birds on the feeding tray—until a Cardinal arrived to join the eating club. There are some scattered Cardinals, Winter and Summer, in our neighborhood, but they are never common and, even if they were, they are too beautiful ever to be considered a bore. The Artist began to boast a bit about his skill as a bird charmer, especially after the Black-capped Chickadees and White-breasted Nuthatches came to dine at the dish of suet he had added to the menu on the feeding tray. He saw a Winter Wren flitting along the rocks that bordered the brook and a Brown Creeper circling up the trunk of a nearby tree like a mechanical mouse, but these birds ignored his feeding tray. The great day, however, came during a real cold spell. Some Goldfinches in their sober Winter coats had been dropping around for an occasional nibble at the grain, and the Artist, hearing a rather musical chatter going on at the feeding tray one day, thought the Goldfinches were back again. He looked through the window and saw a dozen Common Redpolls—visitors from northerly regions, perhaps Labrador—feasting at his outdoor buffet. Then the Artist knew that his feeding station was a success.

That was his first year with it. The second season he had trouble once more in getting it properly started. This time the unwanted first arrival was a Blue Jay. This bold bird, as soon as the tray was set up and the seeds poured out, moved in and practically took sole possession of the feeding station. It ate all the choice seeds and chased away any other bird that ventured near the tray. The Artist went out and drove off the Blue Jay repeatedly. But as soon as the Artist went back to his sketching or painting, the Blue Jay was back in possession of the feeding station again. This kept up until it was a question as to which would collapse first, the Artist or the Blue Jay. The Artist is of tough New England fiber and I am proud to say that it was the Blue Jay that finally became discouraged and quit the place.

One day in early December the Artist and I drove up to his place in the Berkshires. His stone house had been closed tight and the big studio window boarded over since October, and it was like walking into a cold-storage plant when we entered. It seemed colder than it was outdoors and it was 10° Fahrenheit outside. The Artist said that, since we were going to be there only two or three days, it wasn't worth while to turn on the water supply in the house or light the furnace. We could rough it in front of the open fire and carry what water we needed from a spring about a furlong up the road. We really did rough it. We cooked on the open fire and sat in front of it to keep from freezing to death. For three days we wore our caps and big coats indoors as well as outdoors. The Artist slept upstairs and I slept downstairs in front of the open fire. I discovered the first morning that the Artist, still shivering under a bale of blankets, had come down in the night and taken an old bearskin from the floor in front of the hearth and had added that to the pile of blankets under which he was trying to keep warm.

But we did enjoy ourselves when we were out striding over the bleak hills and through the bare woods in search of Winter birds. We found several flocks of Common Redpolls feeding on birch cones and then, about noon of our second day out, the Artist pointed to some birds that were feeding quietly on the seed stalks of a clump of Smooth Sumac. I swung my glasses on them and saw that they were Pine Grosbeaks—five of them—and, from their soft color, probably young of the year. Later that same afternoon, when we were walking in the gloom of a whole hillside of Hemlocks, I said that we might see more birds if we broke out into the open, so we turned toward a gleam of light to the eastward and came to an open patch where, on the ground, we saw a beautiful male Pine Grosbeak wearing the rich raspberry or wine tint that looked as though it had been poured over the bird's head. This bird was quite tame and we moved to within a few feet of it as it fed placidly on some seeds that lay on the hard ground there.

Just before dusk, when we were carrying in wood for the open fire, we saw a Pileated Woodpecker come swooping down over the hillside and over a stone wall to drop into the scrub growth beyond. I walked cautiously to the wall and looked over. There was the Pileated Woodpecker working away on a rotting stump, digging industriously for a supper dish of grubs. This is the prize bird of the area to our eyes, but the Medical Student doesn't believe in its existence. He says it is as mythical as the Roc or the Phoenix. He never has laid eyes on the bird although, on repeated visits to the Artist's stone house on the hillside, he has scoured all that section of the Berkshires in search of the quarry. We never saw a Pileated Woodpecker when the Medical Student was with us, but at other times I saw the bird frequently and the Artist, when painting outdoors in Summer, would see it almost daily. The evidence of the work of the big woodpecker can be seen on the dead trees of the countryside. No other native member of the woodpecker tribe could excavate those big rectangular holes that their strong bills carve through dead wood. But if the Artist and I return from a trip to the Berkshires and tell where and when we saw Pileated Woodpeckers, the Medical Student warns all listeners against us and asserts as Dogberry, that great limb of the law, did in a famous case involving a couple of lying knaves, "Afore God, they are in a tale!"

The last thing we did on this December trip to the Berkshires was to chop down some hickory saplings and cut them into pieces of stove length to be packed away in the car trunk. A farmer in the neighborhood had butchered some pigs and the Artist had purchased one in the raw so that he could smoke it when he returned to the city. Smokehouses were part of the New England landscape where he grew up and he knew how to treat a porker that had been slaughtered and dissected. The law of the Medes and Persians had it that green hickory was the only wood to furnish the best quality smoke for such an operation. So we rolled back to the city with the supply of green hickory for the burning.

But first the hams, shoulders, loins and sides of bacon had to be soaked in a brine of water, brown sugar, salt and saltpeter in proportions that accorded with a formula handed down from the Artist's New England ancestors. He mixed the brine in a tub in his garage and let his booty soak in it for thirty days. Then there came the problem of erecting a smokehouse of some kind outdoors. He couldn't smoke the hams, shoulders, loins and sides of bacon in the garage—which was underneath his house—without smoking a kitchen, a dining room, three bedrooms, two studios and two baths along with the pork products. But the Artist is a man of great invention, a true follower of Leonardo da Vinci. We were walking along the river road one day and, as we crossed a stone bridge over a little ravine, the Artist pointed to something down in the ravine. It was an ash can—or what was left of an ash can. We could see that the bottom was out of it, and there was no lid in sight.

"Treasure trove!" said the Artist joyfully, rubbing his hands in glee. "That ash can will do me for a smokehouse. You take one side and I'll take the other and we'll carry it home."

The wind was cold, the ground was covered with snow, the roads were icy and the way to the Artist's house was all uphill, but we wrestled the ash can up out of the ravine and carried it to its destination, a sheltered nook hard by the Artist's house. There the Artist said as we put down our burden:

"Thanks very much. Your work with this ash can impressed me favorably and if I hear of any other jobs of this kind I'll be glad to recommend you. Come back tomorrow and you'll see an ash can made into a smokehouse. I'll let you see the smoke. For a slight consideration I might even let you smell it."

Out of keen curiosity, I went back the next day and saw what looked like an Indian teepee with smoke seeping out of the top. The Artist was just striding toward the teepee with a bundle of his precious green hickory sticks in his arms.

"Ah, there you are!" he said. "Just pull aside the canvas and peek under."

The canvas wrapping of the teepee was an old striped

awning. A chunk of weather-beaten tarpaulin served as an inner wrapping and a moth-eaten rug was somehow involved in the construction, too. When all this was pulled aside, I could see the ash can resting about a foot off the ground on stone piers and a slow fire of green hickory sticks was giving off a small but steady supply of smoke that went up through the ash can to make its exit at the top of the teepee.

"Close the door of the smokehouse and look up here," said the Artist, pulling aside some of the upper folds of the teepee wrapping.

I looked in as smoke poured out. The Artist had put some steel rods across the top of the ash can and from them he had suspended the hams, shoulders, loins and sides of bacon so that they would be thoroughly smoked. It was an ingenious arrangement. The Artist leaned over to take a few soulful sniffs of the emerging aroma and said as he closed the wrappings of the teepee again:

"Smells wonderful. When it's all smoked, I'll let you taste some."

Later he sent over a slab of bacon as a sample of his skill in handling a smoking operation. The flavor was delicious.

We went on our usual expeditions to the "Lake District" through December. One bleak day we were out on "the peninsula"—a wooded strip of land running out into one of the larger bodies of water—and the Artist, on a side excursion up a hillside through some Scotch Pines, saw a Red Fox trotting silently ahead of him. We put up a Ruffed Grouse that had been feeding on some

255

wild grapes growing along an old stone wall, and further along the wall, where the Bayberry bushes were thick, we saw two Myrtle Warblers. We came to a little open ridge from which we could get a good view of an expanse of open water. There was wind enough to ripple the surface of the lake and it was a few minutes before we noticed a bird of some kind well out in the lake and sitting low in the water. We watched it through our glasses until it turned its head so that we saw the shape of its neck and bill and then we knew it for a Common Loon in its sober Winter plumage. Thoreau wrote of the wild laughter of the Loon on Walden Pond, the "wildest sound that ever is heard here, making the woods ring far and wide." This was a late Loon in our territory. Most of them go through earlier or linger offshore in salt water, but a Loon in December on one of our lakes is a little bit out of season. This bird went along quietly, disappearing beneath the surface at intervals in search of food and coming up so far from its former location that we had quite a search to find it after each dive.

On another part of the big lake we saw Buffleheads, Ring-necked Ducks, American Mergansers and a real "raft" of Black Duck. Black-capped Chickadees, White-breasted Nuthatches and Slate-colored Juncos escorted us along the wood road on our way back over "the peninsula" and, to our surprise, one Red-breasted Nuthatch was foraging on the outer branches of a small White Oak. We do not often find that species on our Winter walks, though some hardy individuals do stay through the cold season in our territory. We had put some miles behind us and, with a bitter wind sweeping down from the north, we were tired and cold and ready to head for home when Herman the Magician announced that we had one more port of call. The Game Warden had phoned him that something interesting had been captured and Herman had promised that we would stop at his house to have a look at it.

We all knew the Game Warden, a man goodly in girth with a jolly disposition to match. He has many virtues and one fault: he hates all hawks with an abiding hatred. They harry the game that is under his protection and his uncompromising attitude is: "Cursed be my tribe if I forgive them!" Beyond that particular prejudice, he is a good friend to man, bird and beast. We knocked on his door and he let out a roar of welcome. Then he led us to the rear of his house where he has a wire enclosure in which he can keep all but the smallest birds or animals taken alive in his bailiwick. A mouse might escape through the mesh, but this was no mouse that he had summoned us to see; it was a Bald Eagle. However, this Bald Eagle was not old enough to be "bald"; the white head and tail feathers that are the *toga virilis* of this species do not appear until the bird's fourth year.

"This one has a busted wing—can't fly far," explained the Game Warden. "It was on the roof of a woman's house when she came back from marketing. She ran inside, locked all the doors and windows and phoned the police to come and take the dangerous bird away. The police called the Society for the Prevention of

257

Cruelty to Animals and they sent over a couple of fellows who caught the bird by shooing it off the roof and grabbing it when it floundered to the ground a little way off. They brought it over here to me. It needs to have that wing treated if it's going to live. Will you fellows take it to the Bronx Zoo? It will get good treatment down there."

We readily agreed to take the bird, provided the Game Warden would bundle it up in some fashion. We were wearing only light woolen gloves on our hands and an eagle's talons are fearsome things. The Game Warden—stout fella!—fearlessly entered the cage, cornered the big bird as it flapped against the wire in an attempt to break out and, with a little help from us, tied the legs and bound down the wings of the captive. I then tucked the trussed Bald Eagle under my arm like a bundle of laundry and we drove down to the Bronx Zoo, where we turned the bird over to our friend George Scott, the chief custodian of the Land Bird House. Scotty—everybody who knows him calls him Scotty— put a splint on the bird's broken wing, but the injury had been too long unattended and infection had set in. The bird died about a week later. While Scotty was doing the best he could for the invalid, he told us of some of his experiences with birds. There was a Purple Finch in a nearby cage and Scotty said:

"Just off a boat—landed on this incoming steamer about fifty miles offshore—a sailor took pity on it and put it in a cage. When they docked, he didn't know what to do with it, so he brought it up here. Nice color—an adult male—but our native wild birds are hard to keep in captivity. Now, look at that cockatoo over there. How old would you say he was?"

We were no experts in judging the age of a cockatoo. We said nothing.

"Well, he's been here thirty years that I know," said Scotty, "and never sick a day in all that time. I don't think you could kill him with an ax. Just give him a mess of sunflower seed and a slice of banana now and then and he stays in great shape. But bring in a native bird and it goes off its feed in a week. In fact, it rarely comes on its feed. Still, we had one Robin that lived four years. It was a cripple—had one bad wing and couldn't fly. A man rescued it after it had been mauled by a cat. Then he heard it was against the law to keep a wild bird in captivity, so he brought it here. One day a man phoned up from a bank on 149th Street and said there was a hummingbird that had been flying around a chandelier in the bank for two hours without stopping—which I doubted—and would I come and get it? So I picked up my fishing pole and started after it."

A fishing pole! That was a new one on us. We never before had heard of a man going fishing for humming-birds!

"Shucks!" said Scotty with a chuckle. "That's the way the collectors get 'em. Just put a twig covered with birdlime on the tip of your fishing pole. You know how hummingbirds fly—they dart here and there and then hover like a bee. Well, when they hover, you just touch them with the tip of your fishing pole and the twig with

259

the birdlime on it sticks to their feathers. Down flutters
the hummingbird and you pick it up. The birdlime can
be washed off easily with kerosene and the hummingbird is as good as new again. They gave me a big step-
ladder at the bank and I had that hummingbird after
only fifteen minutes of fishing. But I made better time
than that when I caught the Golden-fronted Green
Bulbul in Yonkers."

That sounded like something from the imaginative
memoirs of the late Baron Munchausen. We knew for
sure that the Golden-fronted Green Bulbul was not a
permanent resident of the Yonkers area and we were
flabbergasted at the suggestion that it was even a casual
visitor in that region.

"Don't ask me how it got there," said Scotty. "All I
know is that a lady phoned and described this bird and
said it came every day to her dooryard at a certain time.
From the description, I figured it couldn't be anything
else than a Golden-fronted Green Bulbul. We had some
of those birds in a cage here, so I took one as a decoy, put
it in a trapping cage and went over to the Yonkers
dooryard at about the time the lady said it showed up
every day. By golly! I hardly had the decoy set down
and the meal worms spread for bait when around came
the strange bird on schedule—and it was a Golden-
fronted Green Bulbul as advertised. You know, they live
in Burma and Malaya. Well, sir, this free bird came
down to investigate the decoy, then went for the meal
worms—and I closed the trap and was on my way back
to the park in about fifteen minutes all told. I suppose

the bird escaped from somebody's house cage, but nobody ever came around to claim it. Anyway, it was a
real Golden-fronted Green Bulbul that was captured in
Yonkers, which must be an ornithological record of
some kind."

We agreed with Scotty and asked him to take us with
him if he went on any more such record-breaking trips.
We had poor luck ourselves when we took the assign-
ment of covering our ridge and the east bank of the
Hudson for the Christmas census of birds conducted
by the members of the Linnaean Society on a day in
late December. The temperature was only a few degrees below freezing, but a fierce west wind that swept
across the river drove all land birds to cover on our
ridge. We missed half a dozen species that we knew
must be somewhere about the premises. We had seen
them a day earlier and we were to see them the day
after, but they were nowhere in sight the day we took
the census. We couldn't find the Sparrow Hawk, the
Hairy Woodpecker or the Cardinal that the Artist and
I had seen regularly in our morning saunters along the
river road. We didn't see a single Goldfinch—and the
next day I saw a flock of sixty. We didn't even see a
Crow!

We were, in fact, ashamed to hand in our report because we had logged only twenty-one species, which
was quite bad even for the limited territory assigned to
us. We started early in the morning by walking northward along the ridge and then down to the riverbank to
look over the usual concentration of gulls at a sewer

261

outlet. We found about three hundred Herring Gulls there and were grievously disappointed when we couldn't see a single Great Black-backed Gull among them. We generally are able to locate at least one Great Black-backed Gull in any large gathering of that kind. However, the keen eye of the Artist took note of a small gull on the outer fringe of the floating group, and when it took wing we saw that it was a Bonaparte's Gull, which is a rare sight for us at any time along the river and certainly we never expected to see one so late in December.

I spotted in the distance a blackish bird with white on its head and, after we moved up the riverbank to get a closer view, we saw that it was a male Surf Scoter in splendid feather. A freight train that came booming along the track at the edge of the river scared the Surf Scoter into taking wing and it obligingly flew past us to give us a perfect view of its striking black-and-white head pattern before it wheeled off up the river. Then Herman the Magician announced that he had an odd-looking diving duck under scrutiny through his glasses. We took shelter from the wind behind a railroad signal tower to study this bird. It looked dirty and discolored, as though it had blundered into an oil slick somewhere along the water, but eventually we were able to iden-tify it as a bedraggled female Old Squaw.

These were good birds for us to find on the river and, for a few minutes, we didn't mind the cruel wind that drew tears from our eyes as we looked at them through our field glasses, but our luck ran out after that and the only species we logged on a long and bitter cold patrol of riverbank and ridge were House Sparrow, Song Sparrow, White-throated Sparrow, White-breasted Nuthatch, Downy Woodpecker, Slate-colored Junco, Black-capped Chickadee, Blue Jay, Starling, Cedar Waxwing, Black Duck, Mallard, Greater Scaup, American Golden-eye, Cooper's Hawk, Red-tailed Hawk and Ring-necked Pheasant. The one Red-tailed Hawk that we saw was an old friend that was spending its third Winter hunting our ridge. We knew it by the gap in the secondaries of its left wing. A shot or some other injury evidently had cut away part of the flesh there and the feathers never had grown in again. The Cedar Waxwings, nine of them, were huddled glumly in a little Pin Oak near a berried tree of some kind—a cultivated species on a lawn—on which they had been feeding. They had their shoulders hunched against the biting wind and didn't seem any more pleased with the weather than we were.

But we survived worse weather than that and the Astronomer, who is our official photographer, has pictures of our group that he could pass off as scenes from some Arctic expedition. Once when we had fought our way across an open plain through a whirling snowstorm, we gained the comparative quiet of a patch of woods and were astonished to see dark insects fluttering against the white background of the fallen snow. They turned out to be small brownish moths that the wind had blown out of the surrounding trees to which they had been clinging after hatching from the egg masses attached to the bark. I went down on my hands and knees in the snow to study the moths under my pocket magnifying glass and, when I looked them up later in *The Moth Book*, I found that they belonged to the family of my ancient enemies, the *Geometridae*, the "inch-worms" or "loopers" that ravage the shade trees of our territory. This species is called the Fall Canker-worm because the moths—or some of them, at least—hatch out in Autumn, and thus it differs from another but no more amiable member of the family, the Spring Canker-worm, the moths of which species have enough sense to wait until Winter has passed before emerging to take wing.

We plodded on through the snow that was deep underfoot and still falling heavily around us in the woods. Rain is wicked to face in Winter, but snow is wonderful in the wild. Often I look through my magnifying glass at single flakes that I catch on my sleeve or my glove so that I can gloat over the amazing beauty of the ever-changing designs with which these delicate crystalline masterpieces of Nature are fashioned—

> So *purely, so palely,*
> *Tinily, surely,*
> *Mightily, frailly,*
> *Insculped and embossed,*
> *With His hammer of wind,*
> *And His graver of frost.*

18

And so, without more circumstance at all,
I hold it fit that we shake hands and part.
HAMLET, ACT I, Sc. 5

Well, that's the way of it with us in the open. We have
made the full circuit of a rambling year from the snows
of January through the wind-swept days of March,
through the glorious greenery of May, through the hush
of Summer on the Berkshire hills. We saw the surf
breaking on the Fire Island beach and we heard the
sharp cries of the graceful terns in unwearied flight
offshore. We stood watch on high places for migrating
hawks when Autumn had colored the landscape around
us and below us. We were thankful for Indian Summer,
the last warm gesture of a dying year, and we plodded
contentedly ahead to face the first snowfall of another
Winter.

Now that this printed pilgrimage is over, I look back
over the pages and realize that I have utterly failed to

paint the picture as we saw it in Nature. All our walks were great fun and many of these chapters are stodgy and dull. Benedick the married man could well say of my tale: "Why, that's spoken like an honest drover; so they sell bullocks." I let Herman the Magician read the manuscript and he wrote with justifiable indignation:

"Where is *Dryopteris hexagonoptera* and where—oh, where!—is that delightful rarity, the Maiden-hair Spleenwort that we found in the rocky crevasse near the Shining Club-moss? What about the Columbine, the Dutchman's-breeches, the Downy Rattlesnake Plantain (a beautiful orchid) and *Viola rotundifolia?* And what about Ginseng, the roots of which have been exported to China for oriental magic purposes these many years? What about the hillside we found covered with Blue Cohosh—and the same hillside the only place in our area for *Liparis liliifolia?*"

Alas, what about so many things that I have overlooked, omitted or marred in the telling? For one thing, I drew much from Nature in secret that I am reluctant to acknowledge in public. Shall a sober citizen break down and confess how deeply he is stirred by such foolish things as "the rift of dawn, the reddening of the rose"? Shall the resident of a respectable suburban neighborhood admit that often, of a clear cold Winter night, he stood gazing in speechless admiration at the stark poetry of "the black elm tops 'mong the freezing stars" or "the long glories of the Winter moon" on a snow-covered landscape? Truly "I triumphed and I saddened with all weather," but it was Francis Thompson who had the courage to put such an intimate revelation in words. Such things are beyond me.

In Ecclesiastes it is written: "And one shall rise up at the voice of a bird." Yea, more than one; all of our strolling company will answer such a luring call through the calendar year, for birds have been the primary objects of our searches on most days in the field. But we found many other things of interest along the road from dawn to dusk by hill and hollow. We peered at little insects

of strange shapes and odd colors that we uncovered when we turned over flat stones in meadows, and we looked at those immense whirling electrical storms called "sun spots"—some of them 25,000 miles in diameter and all of them 93,000,000 miles away—through the Astronomer's darkened glasses. We watched the migrations of the birds and we took note of the march of the constellations across the night sky as the seasons changed. We looked closely at the stamens and pistils of flowers through magnifying glasses and we peered distantly at the Great Nebula in Andromeda through a telescope. All this cost us little and we enjoyed it greatly. It has left us with happy memories and high hopes. Thoreau reminds us that "there is more day to dawn." Tomorrow and tomorrow and tomorrow we shall be on the alert for "the earliest pipe of half-awakened birds."

Appendix

This is not by any means a complete list of the birds, trees, flowers, mammals and insects that we saw on our walks. It is merely a list of the species mentioned in the text, with the scientific names added for the benefit of those readers who may wish to know the particular species or subspecies that we found in our territory. Authorities often differ with regard to common or scientific names of genera or species. The only thing for an author to do in a book like this is to choose some recognized authority in each field and stick to it throughout. Even so, there is the further difficulty of occasional (or even periodical!) revisions in nomenclature. The Committee on Nomenclature of the American Ornithologists' Union has been revising the list of the common and scientific names of the birds of North America and already some of the bird names offered here are out of date, but since the revised A.O.U. Check-List has not been completed and the ordinary reader has no ready reference for changes made or to come, it was decided to use the nomenclature of the latest edition of Chapman's *Handbook of Birds of Eastern North America,* which is readily available to all bird students. A similar difficulty was encountered in the botanical field and a similar decision made. The reference volumes quoted here (Britton and Brown, Second Edition) are being revised and some names are being changed in the process, but it will be some time before the extensive task is completed, wherefor it was decided to stick to the Second Edition throughout.

Birds

BALDPATE. *See* Duck, Baldpate.
BITTERN, AMERICAN. *Botaurus lentiginosus.*
   EASTERN LEAST. *Ixobrychus exilis exilis.*
BLACKBIRD, EASTERN RED-WINGED. *Agelaius phoeniceus phoe-*
   niceus.
   RUSTY. *Euphagus carolinus.*
BLUEBIRD, EASTERN. *Sialia sialis sialis.*
BOBOLINK. *Dolichonyx oryzivorus.*
BOB-WHITE. *Colinus virginianus virginianus.*
BUFFLEHEAD. *See* Duck, Bufflehead.
BUNTING, SNOW. *Plectrophenax nivalis nivalis.*
CANVASBACK. *See* Duck, Canvasback.
CARDINAL, EASTERN. *Richmondena cardinalis cardinalis.*
CATBIRD. *Dumatella carolinensis.*
CHAT, YELLOW-BREASTED. *Icteria virens virens.*
CHEBEC. *See* Flycatcher, Least.
CHEWINK. *See* Towhee, Red-eyed.
CHICKADEE, BLACK-CAPPED. *Penthestes atricapillus atricapillus.*
COOT, AMERICAN. *Fulica americana americana.*
CORMORANT, DOUBLE-CRESTED. *Phalacrocorax auritus auritus.*
   EUROPEAN. *Phalacrocorax carbo carbo.*
COWBIRD. *Molothrus ater ater.*
CREEPER, BROWN. *Certhia familiaris americana.*
CROW, EASTERN. *Corvus brachyrhynchos brachyrhynchos.*
CUCKOO, BLACK-BILLED. *Coccyzus erythrophthalmus.*
   YELLOW-BILLED. *Coccyzus americanus americanus.*
CURLEW, HUDSONIAN. *Phaeopus hudsonicus.*
DOVE, EASTERN MOURNING. *Zenaidura macroura carolinensis.*
DOWITCHER, EASTERN. *Limnodromus griseus griseus.*
DUCK, AMERICAN GOLDEN-EYE. *Glaucionetta clangula americana.*
   BALDPATE. *Mareca americana.*
   BUFFLEHEAD. *Charitonetta albeola.*
   CANVASBACK. *Nyroca valisineria.*
   COMMON BLACK. *Anas rubripes tristis.*
   COMMON MALLARD. *Anas platyrhynchos platyrhynchos.*
   EUROPEAN WIDGEON. *Mareca penelope.*
   GADWALL. *Chaulelasmus streperus.*

OLD SQUAW. *Clangula hyemalis.*
PINTAIL, AMERICAN. *Dafila acuta tzitzihoa.*
REDHEAD. *Nyroca americana.*
RING-NECKED. *Nyroca collaris.*
RUDDY. *Erismatura jamaicensis rubida.*
SCAUP, GREATER. *Nyroca marila.*
SCAUP, LESSER. *Nyroca affinis.*
SHOVELLER. *Spatula clypeata.*
TEAL, BLUE-WINGED. *Querquedula discors.*
TEAL, EUROPEAN. *Nettion crecca.*
TEAL, GREEN-WINGED. *Nettion carolinense.*
WOOD. *Aix sponsa.*
EAGLE, AMERICAN GOLDEN. *Aquila chrysaetos canadensis.*
SOUTHERN BALD. *Haliaeetus leucocephalus leucocephalus.*
FINCH, EASTERN PURPLE. *Carpodacus purpureus purpureus.*
FLICKER, NORTHERN. *Colaptes auratus luteus.*
FLYCATCHER, ACADIAN. *Empidonax virescens.*
ALDER. *Empidonax trailli trailli.*
LEAST. *Empidonax minimus.*
NORTHERN CRESTED. *Myiarchus crinitus boreus.*
OLIVE-SIDED. *Nuttallornis mesoleucus.*
YELLOW-BELLIED. *Empidonax flaviventris.*
GADWALL. *See* Duck, Gadwall.
GALLINULE, FLORIDA. *Gallinula chloropus cachinnans.*
GOLDEN-EYE. *See* Duck, Golden-eye.
GOLDFINCH, EASTERN. *Spinus tristis tristis.*
GOOSE, CANADA. *Branta canadensis canadensis.*
GRACKLE, PURPLE. *Quiscalus quiscula quiscula.*
GREBE, HORNED. *Colymbus auritus.*
PIED-BILLED. *Podilymbus podiceps podiceps.*
GROSBEAK, EASTERN EVENING. *Hesperiphona vespertina vespertina.*
PINE. *Pinicola enucleator leucura.*
ROSE-BREASTED. *Hedymeles ludovicianus.*
GROUSE, EASTERN RUFFED. *Bonasa umbellus umbellus.*
GULL, BONAPARTE'S. *Larus philadelphia.*
GREAT BLACK-BACKED. *Larus marinus.*
HERRING. *Larus argentatus smithsonianus.*
LAUGHING. *Larus atricilla.*
RING-BILLED. *Larus delawarensis.*
HAWK, AMERICAN ROUGH-LEGGED. *Buteo lagopus sancti-johannis.*
BROAD-WINGED. *Buteo platypterus platypterus.*

271

COOPER'S. *Accipiter cooperi.*

DUCK. *Falco peregrinus anatum.*

EASTERN RED-TAILED. *Buteo borealis borealis.*

EASTERN SPARROW. *Falco sparverius sparverius.*

MARSH. *Circus hudsonius.*

NORTHERN RED-SHOULDERED. *Buteo lineatus lineatus.*

PIGEON. *Falco columbarius columbarius.*

SHARP-SHINNED. *Accipiter velox velox.*

HERON, BLACK-CROWNED NIGHT. *Nycticorax nycticorax hoactli.*

EASTERN GREEN. *Butorides virescens virescens.*

GREAT BLUE. *Ardea herodias herodias.*

HUMMINGBIRD, RUBY-THROATED. *Archilochus colubris.*

JAY, BLUE. *Cyanocitta cristata cristata.*

JUNCO, SLATE-COLORED. *Junco hyemalis hyemalis.*

KILLDEER. *Oxyechus vociferus vociferus.*

KINGBIRD. *Tyrannus tyrannus.*

KINGFISHER, EASTERN BELTED. *Megaceryle alcyon alcyon.*

KINGLET, EASTERN RUBY-CROWNED. *Corthylio calendula calendula.*

GOLDEN-CROWNED. *Regulus satrapa satrapa.*

LARK, NORTHERN HORNED. *Otocoris alpestris alpestris.*

PRAIRIE HORNED. *Otocoris alpestris praticola.*

LOON, COMMON. *Gavia immer immer.*

MALLARD. *See* Duck, Common Mallard.

MARTIN, PURPLE. *Progne subis subis.*

MEADOWLARK, EASTERN. *Sturnella magna magna.*

MERGANSER, AMERICAN. *Mergus merganser americanus.*

HOODED. *Lophodytes cucullatus.*

RED-BREASTED. *Mergus serrator.*

NIGHTHAWK, EASTERN. *Chordeiles minor minor.*

NUTHATCH, NORTHERN WHITE-BREASTED. *Sitta carolinensis carolinensis.*

RED-BREASTED. *Sitta canadensis.*

OLD SQUAW. *See* Duck, Old Squaw.

ORIOLE, BALTIMORE. *Icterus galbula.*

ORCHARD. *Icterus spurius.*

OSPREY. *Pandion haliaetus carolinensis.*

OVEN-BIRD. *Seiurus aurocapillus.*

OWL, EASTERN SCREECH. *Otus asio naevius.*

GREAT HORNED. *Bubo virginianus virginianus.*

NORTHERN BARRED. *Strix varia varia.*

SAW-WHET. *Cryptoglaux acadica acadica.*
SNOWY. *Nyctea nyctea.*
PEWEE, EASTERN WOOD. *Myiochanes virens.*
PHEASANT, RING-NECKED. *Phasianus colchicus torquatus.*
PHOEBE, EASTERN. *Sayornis phoebe.*
PINTAIL. *See* Duck, Pintail.
PLOVER, BLACK-BELLIED. *Squatarola squatarola.*
  PIPING. *Charadrius melodus.*
  SEMIPALMATED. *Charadrius semipalmatus.*
RAIL, KING. *Rallus elegans elegans.*
  SORA. *Porzana carolina.*
  VIRGINIA. *Rallus limicola limicola.*
REDHEAD. *See* Duck, Redhead.
REDPOLL, COMMON. *Acanthis linaria linaria.*
REDSTART, AMERICAN. *Setophaga ruticilla.*
ROBIN, EASTERN. *Turdus migratorius migratorius.*
SANDERLING. *Crocethia alba.*
SANDPIPER, LEAST. *Pisobia minutilla.*
  SEMIPALMATED. *Ereunetes pusillus.*
  SPOTTED. *Actitus macularia.*
SAPSUCKER, YELLOW-BELLIED. *Sphyrapicus varius varius.*
SCAUP. *See* Duck, Scaup.
SCOTER, AMERICAN. *Oidemia americana.*
  SURF. *Melanitta perspicillata.*
  WHITE-WINGED. *Melanitta deglandi.*
SHOVELLER. *See* Duck, Shoveller.
SHRIKE, NORTHERN. *Lanius borealis borealis.*
SISKIN, PINE. *Spinus pinus pinus.*
SNIPE, WILSON'S. *Capella delicata.*
SORA. *See* Rail, Sora.
SPARROW, EASTERN CHIPPING. *Spizella passerina passerina.*
  EASTERN GRASSHOPPER. *Ammodramus savannarum australis.*
  EASTERN SAVANNAH. *Passerculus sandwichensis savanna.*
  EASTERN SONG. *Melospiza melodia melodia.*
  EASTERN TREE. *Spizella arborea arborea.*
  EASTERN VESPER. *Pooecetes gramineus gramineus.*
  FIELD. *Spizella pusilla pusilla.*
  FOX. *Passerella iliaca iliaca.*
  HOUSE OR ENGLISH. *Passer domesticus domesticus.*
  SHARP-TAILED. *Ammospiza caudacuta caudacuta.*
  SWAMP. *Melospiza georgiana.*
  WHITE-THROATED. *Zonotrichia albicollis.*

STARLING. *Sturnus vulgaris vulgaris.*
SWALLOW, BANK. *Riparia riparia riparia.*
   BARN. *Hirundo erythrogaster.*
   NORTHERN CLIFF. *Petrochelidon albifrons albifrons.*
   ROUGH-WINGED. *Stelgidopteryx ruficollis serripennis.*
   TREE. *Iridoprocne bicolor.*
SWAN, MUTE. *Sthenelides olor.*
SWIFT, CHIMNEY. *Chaetura pelagica.*
TANAGER, SCARLET. *Piranga erythromelas.*
TEAL. *See* Duck, Teal.
TERN, COMMON. *Sterna hirundo hirundo.*
   LEAST. *Sterna antillarum antillarum.*
THRASHER, BROWN. *Toxostoma rufum.*
THRUSH, EASTERN HERMIT. *Hylocichla guttata faxoni.*
   OLIVE-BACKED. *Hylocichla ustulata swainsoni.*
   WILSON'S. *See* Veery.
   WOOD. *Hylocichla mustelina.*
TOWHEE, RED-EYED. *Pipilo erythrophthalmus erythrophthalmus.*
TURNSTONE, RUDDY. *Arenaria interpres morinella.*
VEERY. *Hylocichla fuscescens fuscescens.*
VIREO, BLUE-HEADED. *Vireo solitarius solitarius.*
   EASTERN WARBLING. *Vireo gilvus gilvus.*
   RED-EYED. *Vireo olivaceus.*
   WHITE-EYED. *Vireo griseus griseus.*
   YELLOW-THROATED. *Vireo flavifrons.*
VULTURE, TURKEY. *Cathartes aura septentrionalis.*
WARBLER, BAY-BREASTED. *Dendroica castanea.*
   BLACK AND WHITE. *Mniotilta varia.*
   BLACKBURNIAN. *Dendroica fusca.*
   BLACK-POLL. *Dendroica striata.*
   BLACK-THROATED BLUE. *Dendroica caerulescens caerulescens.*
   BLACK-THROATED GREEN. *Dendroica virens virens.*
   BLUE-WINGED. *Vermivora pinus.*
   CANADA. *Wilsonia canadensis.*
   CHESTNUT-SIDED. *Dendroica pensylvanica.*
   EASTERN YELLOW. *Dendroica aestiva aestiva.*
   HOODED. *Wilsonia citrina.*
   MAGNOLIA. *Dendroica magnolia.*
   MYRTLE. *Dendroica coronata.*
   NASHVILLE. *Vermivora ruficapilla ruficapilla.*
   NORTHERN PARULA. *Compsothlypis americana pusilla.*
   NORTHERN PINE. *Dendroica pinus pinus.*

274

PRAIRIE. *Dendroica discolor discolor.*
WILSON'S. *Wilsonia pusilla pusilla.*
WORM-EATING. *Helmitheros vermivorus.*
YELLOW PALM. *Dendroica palmarum hypochrysea.*
WATER-THRUSH, LOUISIANA. *Seiurus motacilla.*
  NORTHERN. *Seiurus noveboracensis noveboracensis.*
WAXWING, CEDAR. *Bombycilla cedrorum.*
WHIP-POOR-WILL, EASTERN. *Antrostomus vociferus vociferus.*
WIDGEON. *See* Duck, European Widgeon.
WILLET, WESTERN. *Catoptrophorus semipalmatus inornatus.*
WOODCOCK. *Philohela minor.*
WOODPECKER, EASTERN HAIRY. *Dryobates villosus villosus.*
  GOLDEN-WINGED. *See* Flicker.
  NORTHERN DOWNY. *Dryobates pubescens medianus.*
  NORTHERN PILEATED. *Ceophloeus pileatus abieticola.*
WREN, EASTERN WINTER. *Nannus hiemalis hiemalis.*
  HOUSE. *Troglodytes aedon aedon.*
  LONG-BILLED MARSH. *Telmatodytes palustris palustris.*
YELLOW-LEGS, GREATER. *Totanus melanoleucus.*
  LESSER. *Totanus flavipes.*
YELLOW-THROAT, NORTHERN. *Geothlypis trichas brachidactyla.*

Insects (Except Lepidoptera)

ANT-LION. *Myrmeleon sp.*
BEETLE, JAPANESE. *Popillia japonica.*
CANKER-WORM, FALL. *Alsophila pometaria.*
  SPRING. *Paleacrita vernata.*
CATERPILLAR, TENT. *Malacosoma americana.*
CICADA-KILLER. *Sphecius speciosus.*
CRICKET, COMMON. *Gryllus assimilis.*
ELATOR, EYED. *Alaus oculatus.*
LOCUST, SEVENTEEN-YEAR. *Magicicada septendecim.*

Lepidoptera (Butterflies and Moths)

ADMIRAL, RED. *Vanessa atalanta.*
BLUE, COMMON. *Lycaena pseudargiolus.*

275

BUCKEYE. *Junonia coenia.*
FRITILLARY, GREAT SPANGLED. *Argynnis cybele.*
MONARCH. *Danais plexippus.*
MOTH, CECROPIA. *Samia cecropia.*
MOURNING-CLOAK. *Aglais antiopa.*
SKIPPER, SILVER-SPOTTED. *Epargyreus tityrus.*
SWALLOWTAIL, TIGER. *Papilio turnus.*
    SPICE-BUSH. *Papilio troilus.*
VICEROY. *Basilarchia archippus.*
WOOD-SATYR, LITTLE. *Euptychia euryta.*

Mammals

BEAR, AMERICAN BLACK. *Ursus americana americana.*
BEAVER, CANADA. *Castor canadensis canadensis.*
DEER, WHITE-TAILED. *Odocoileus americanus americanus.*
FISHER (PENNANT'S MARTEN) *Martes pennanti.*
FOX, EASTERN GRAY. *Urocyon cinereoargenteus cinereoargenteus.*
    EASTERN RED. *Vulpes fulva.*
MINK, AMERICAN. *Mustela vison.*
MUSK-RAT. *Ondatra zibethica zibethica.*
OTTER, CANADA. *Lutra canadensis.*
PORCUPINE, CANADA. *Erethizon dorsatum dorsatum.*
RABBIT, NEW ENGLAND COTTONTAIL. *Sylvilagus transitionalis.*
SKUNK, NEW ENGLAND STRIPED. *Mephitis putida.*
SQUIRREL, EASTERN GRAY. *Sciurus carolinensis carolinensis.*
WOODCHUCK, EASTERN. *Marmota monax monax.*

Plants (Trees and Flowers)

ALDER, BLACK. *Ilex verticillata.*
ANEMONE, RUE. *Syndesmon thalictroides.*
ARBUTUS, TRAILING. *Epigaea repens.*
ARROW-WOOD. *Viburnum dentatum.*
ASH, WHITE. *Fraxinus americana.*
ASTER, NEW ENGLAND. *Aster novae-angliae.*
    SAVORY-LEAVED. *Ionactis linariifolius.*

Azalea, Pink. *Azalea nudiflora.*
Baneberry, White. *Actaea alba.*
Bayberry. *Myrica carolinensis.*
Bean, Wild. *Glycine Apios.*
Bearberry. *Uva-Ursi Uva-Ursi.*
Beech. *Fagus grandifolia.*
Beggar-ticks. *Bidens frondosa.*
Birch, Black. *Betula lenta.*
Blackberry, Low Running. *Rubus procumbens.*
Bloodroot. *Sanguinaria canadensis.*
Blueberry, High-bush. *Vaccinium corymbosum.*
Boneset, Climbing. *Mikania scandens.*
Bouncing Bet. *Saponaria officinalis.*
Butter-and-Eggs. *Linaria linaria.*
Button-bush. *Cephalanthus occidentalis.*
Carrot, Wild. *Daucus Carota.*
Catbrier. *Smilax rotundifolia.*
Cat-tail, Broad-leaved. *Typha latifolia.*
    Narrow-leaved. *Typha angustifolia.*
Cedar, Red. *Juniperus virginiana.*
Chestnut. *Castanea dentata.*
Chicory. *Chichorium Intybus.*
Clover, Red. *Trifolium pratense.*
Club-moss, Shining. *Lycopodium lucidulum.*
Cohosh, Black. *Cimicifuga racemosa.*
    Blue. *Caulophyllum thalictroides.*
Coltsfoot. *Petasites Petasites.*
Columbine, Wild. *Aquilegia canadensis.*
Corydalis, Pale. *Capnoides sempervirens.*
Daisy, Common or White. *Chrysanthemum Leucanthemum.*
Dandelion. *Leontodon Taraxacum.*
Dock, Curled. *Rumex crispus.*
Dogwood, Flowering. *Cynoxylon floridum.*
    Panicled. *Cornus femina.*
Dutchman's-breeches. *Bicuculla Cucullaria.*
Elm, American. *Ulmus americana.*
    Slippery. *Ulmus fulva.*
Enchanter's Nightshade. *Circaea lutetiana.*
Evening-Primrose, Common. *Oenothera biennis.*
Fern, American Shield. *Dryopteris intermedia.*
    Christmas. *Polystichum acrostichoides.*
    Cinnamon. *Osmunda cinnamomea.*

INTERRUPTED. *Osmunda Claytoniana.*
MAIDEN-HAIR. *Adiantum pedatum.*
MARGINAL SHIELD. *Dryopteris marginalis.*
OSTRICH. *Matteuccia Struthiopteris.*
SENSITIVE. *Onoclea sensibilis.*
FOXGLOVE, DOWNY FALSE. *Dasystoma flava.*
GARLIC, WILD. *Allium vineale.*
GENTIAN, FRINGED. *Gentiana crinita.*
GERANIUM, WILD. *Geranium maculatum.*
GINSENG. *Panax quinquefolium.*
GOLDENROD, ROCK. *Solidago canadensis.*
GROUND-PINE. *Lycopodium complanatum.*
GUM, SOUR. *Nyssa sylvatica.*
  SWEET. *Liquidambar Styraciflua.*
HAZEL-NUT. *Corylus americana.*
HEMLOCK. *Tsuga canadensis.*
HEPATICA. *Hepatica hepatica.*
HOLLY, AMERICAN. *Ilex opaca.*
HORSE CHESTNUT. *Aesculus Hippocastanum.*
JOE-PYE WEED. *Eupatorium purpureum.*
JUNIPER. *Juniperus communis.*
LADIES'-TRESSES, NODDING. *Ibidium cernuum.*
LADY'S-SLIPPER, PINK. *Fissipes acaulis.*
  YELLOW. *Cypripedium parviflorum.*
LAMB'S QUARTERS. *Chenopodium album.*
LAUREL, MOUNTAIN. *Kalmia latifolia.*
LEEK, WILD. *Allium tricoccum.*
LETTUCE, WILD. *Lactuca canadensis.*
LINDEN, AMERICAN. *Tilia americana.*
LOCUST, BLACK. *Robinia Pseudo-Acacia.*
MAPLE, NORWAY. *Acer platanoides.*
  RED. *Acer rubrum.*
  SUGAR. *Acer saccharum.*
  SYCAMORE. *Acer Pseudo-Platanus.*
MEADOW-SWEET. *Spiraea latifolia.*
MOOSEWOOD. *Acer pensylvanicum.*
MOTHERWORT. *Leonurus cardiaca.*
OAK, BLACK. *Quercus velutina.*
  MOSSY-CUP. *Quercus macrocarpa.*
  PIN. *Quercus palustris.*
  RED. *Quercus rubra.*
  ROCK CHESTNUT. *Quercus prinus.*

278

Swamp White. *Quercus bicolor.*
White. *Quercus alba.*
Orchis, Showy. *Galeorchis spectabilis.*
Partridge-berry. *Mitchella repens.*
Peanut, Hog. *Falcata comosa.*
Pickerel Weed. *Pontedaria cordata.*
Pine, Pitch. *Pinus rigida.*
  Scotch. *Pinus sylvestris.*
  White. *Pinus Strobus.*
Pogonia, Rose. *Pogonia ophioglossoides.*
Poison Ivy. *Toxicodendron radicans.*
Polypody, Common. *Polypodium vulgare.*
Ragweed, Great. *Ambrosia trifida.*
Raspberry, Wild Red. *Rubus strigosus.*
Rattlesnake Plantain, Downy. *Peramium pubescens.*
Sassafras. *Sassafras Sassafras.*
Shad-bush. *Amelanchior intermedia.*
Skunk Cabbage. *Spathyema foetida.*
Snake-root, White. *Eupatorium urticaefolium.*
Spice-bush. *Benzoin aestivale.*
Spleenwort, Ebony. *Asplenium platyneuron.*
  Maiden-hair. *Asplenium Trichomanes.*
Spring Beauty, Narrow-leaved. *Claytonia virginica.*
Sumac, Smooth Upland. *Rhus glabra.*
  Staghorn. *Rhus hirta.*
Sweet Pepper-bush. *Clethra alnifolia.*
Sycamore. *Platanus occidentalis.*
Trillium, Purple. *Trillium erectum.*
Tulip Tree. *Liriodendron Tulipifera.*
Tupelo. *See* Gum, Sour.
Violet, Dog's-tooth. *Erythronium americanum.*
Virginia Creeper. *Parthenocissus quinquefolia.*
Wafer-ash. *Ptelea trifoliata.*
Willow, Pussy. *Salix discolor.*
Wind-flower. *Anemone quinquefolia.*
Witch Hazel. *Hamamelis virginiana.*
Yam-root, Wild. *Dioscorea villosa.*
Yarrow. *Achillea Millefolium.*
Yew, American. *Taxus canadensis.*

Know what you see!

Three years ago, spurred by the example of Leslie P. Thompson, I began to keep a fishing diary. My first concern was to note the size and species of the monsters brought to my net, but as my interest deepened I began to make note of the time and the climatic conditions, and if I was after sea fish, of the tide; I tried to calculate the underwater architecture. There is in all of us an impulse to take note of what we love best in nature, but unless impulse teams up with habit, nothing gets written.

John Kieran's *Footnotes on Nature* is a book of the seasons, enriched by recollection, of a city dweller who refreshes his spirit by walking and by bird-watching in the woods and fields of New York and New England. Most of us have come to know Mr. Kieran as that delightfully human encyclopedia on "Information Please." In this pleasantly confiding book he shows us the origin and orderliness of his knowledge. When he was eleven his father bought a forty-acre farm in Dutchess County, New York; there, during the long summer vacations, young John acquired his early knowledge of trees and birds, the wild flowers and the streams, and there as a young schoolteacher (to me a significant admission) his antenna reached out for a closer and closer observation. The "want to know" was natural in the boy as it was persistent in the man; and being blessed with clear eyes and a fine memory, Mr. Kieran has been able to gratify his insatiable curiosity about our flora and fauna. Often in the company of a friendly quartet, the Dramatic Critic (Brooks Atkinson), the Artist, Herman the Magician ("He loves to catch fish, eat fish and talk about fish"), and the Astronomer, he has walked down the years on the

[1] The conscientious few might like to know that the Massachusetts Park and Forest Association, 3 Joy Street, Boston, is waging a good fight and needs contributions.